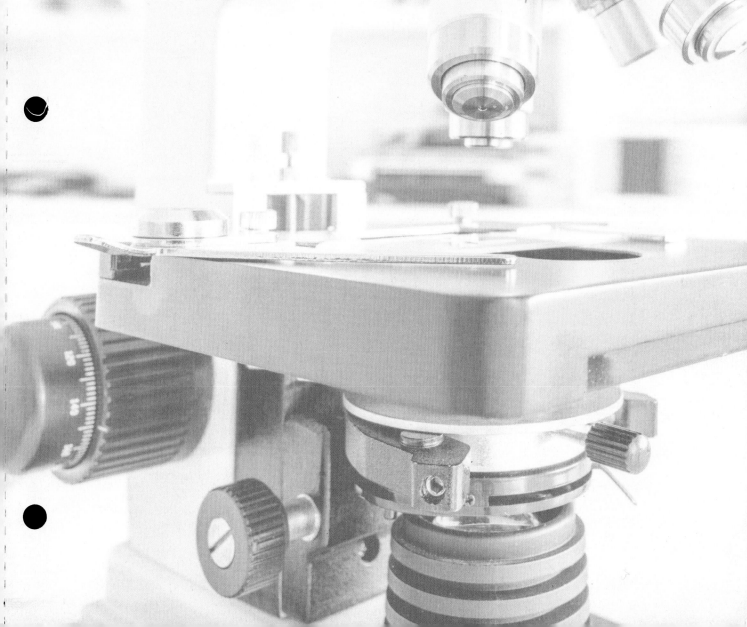

# LABORATORY MANUAL

## Microbiology
# FUNDAMENTALS
### A Clinical Approach

D1298805

# LABORATORY MANUAL

# Microbiology
# FUNDAMENTALS
## A Clinical Approach
### Third Edition

## Steven Obenauf
*Broward College*

## Susan Finazzo
*Perimeter College, Georgia State University*

Mc
Graw
Hill
Education

LABORATORY MANUAL FOR MICROBIOLOGY FUNDAMENTALS: A CLINICAL APPROACH, THIRD EDITION

4 5 6 7 8 9 LMN 21 20 19

ISBN 978-1-260-16348-3
MHID 1-260-16348-2

Senior Portfolio Manager: *Marjia Magner*
Product Developer: *Darlene M. Schueller*
Marketing Manager: *Valerie L. Kramer*
Content Project Managers: *Mary Jane Lampe/Christina Nelson*
Buyer: *Laura Fuller*
Design: *Matt Backhaus*
Content Licensing Specialist: *Lori Hancock*
Cover Image: *(Green agar) ©Jimgames/Getty Images RF; (Scientist) ©Monty Rakusen/Getty Images; (Surgeon) ©ERproductions Ltd/Blend Images LLC; (SEM of Ebola virus) Source: CDC/National Institute of Allergy and Infectious Diseases (NIAID); Microscope: ©science photo/Shutterstock RF*
Compositor: *Aptara®, Inc.*

# About the Authors

**Steven Obenauf,** Ph.D., has been teaching microbiology lectures and laboratories at Broward College since 1993. For six years before moving to the Fort Lauderdale area, he taught medical immunology and medical virology to first- and second-year medical students as well as first-year pharmacy students. His dedication to his students has led him to be a recipient of both Professor of the Year and Endowed Teaching Chair awards for undergraduate teaching and the "Golden Apple" teaching award bestowed by the graduating medical class. Dr. Obenauf is a registered microbiologist and belongs to both the American Society for Microbiology (ASM) and National Association of Biology Teachers. For 10 years, he served as the secretary and newsletter editor for the Florida branch of ASM. He loves using various forms of technology as part of his teaching and has given workshops and mentored other faculty at his college on in-class and online technology tools. Dr. Obenauf is currently Professor of Biological Sciences at Broward College. This manual brings together two of his passions: microbiology and photography.

Courtesy of Gail Obenauf

**Susan Finazzo,** M.S., Ph.D., has a diverse academic background with degrees in biology, geology, microbiology, and horticulture. Her background, clinical work experience, and innate interest in the living world have allowed her to teach a broad variety of subjects. Her first academic love has always been microbiology. Throughout the past 20 years, Dr. Finazzo has taught students at both the two-year college level and university level. She is a member of the National Association of Biology Teachers and has held several leadership positions including serving as president in 2017. She has been nominated for several teaching and service awards, and has twice been the recipient of an Endowed Teaching Award. Dr. Finazzo is very interested in new pedagogical approaches and tools to address the learning styles of the millennial student.

©Susan F. Finazzo

# Contents

McGraw-Hill Connect® is a highly reliable, easy-to-use homework and learning management solution that utilizes learning science and award-winning adaptive tools to improve student results.

# Homework and Adaptive Learning

- Connect's assignments help students contextualize what they've learned through application, so they can better understand the material and think critically.
- Connect will create a personalized study path customized to individual student needs through SmartBook®.
- SmartBook helps students study more efficiently by delivering an interactive reading experience through adaptive highlighting and review.

## Connect's Impact on Retention Rates, Pass Rates, and Average Exam Scores

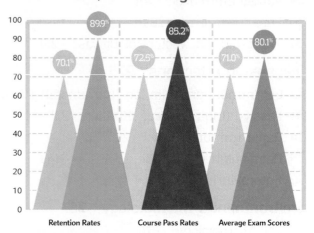

without Connect    with Connect

Using **Connect** improves retention rates by **19.8%**, passing rates by **12.7%, and** exam scores by **9.1%.**

Over **7 billion questions** have been answered, making McGraw-Hill Education products more intelligent, reliable, and precise.

73% of instructors who use **Connect** require it; instructor satisfaction **increases** by 28% when **Connect** is required.

# Quality Content and Learning Resources

- Connect content is authored by the world's best subject matter experts, and is available to your class through a simple and intuitive interface.
- The Connect eBook makes it easy for students to access their reading material on smartphones and tablets. They can study on the go and don't need internet access to use the eBook as a reference, with full functionality.
- Multimedia content such as videos, simulations, and games drive student engagement and critical thinking skills.

# Robust Analytics and Reporting

©Hero Images/Getty Images

- Connect Insight® generates easy-to-read reports on individual students, the class as a whole, and on specific assignments.

- The Connect Insight dashboard delivers data on performance, study behavior, and effort. Instructors can quickly identify students who struggle and focus on material that the class has yet to master.

- Connect automatically grades assignments and quizzes, providing easy-to-read reports on individual and class performance.

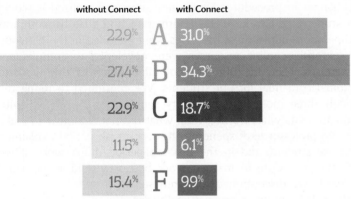

## Impact on Final Course Grade Distribution

| without Connect | | with Connect |
|---|---|---|
| 22.9% | A | 31.0% |
| 27.4% | B | 34.3% |
| 22.9% | C | 18.7% |
| 11.5% | D | 6.1% |
| 15.4% | F | 9.9% |

More students earn **As** and **Bs** when they use **Connect**.

# Trusted Service and Support

- Connect integrates with your LMS to provide single sign-on and automatic syncing of grades. Integration with Blackboard®, D2L®, and Canvas also provides automatic syncing of the course calendar and assignment-level linking.

- Connect offers comprehensive service, support, and training throughout every phase of your implementation.

- If you're looking for some guidance on how to use Connect, or want to learn tips and tricks from super users, you can find tutorials as you work. Our Digital Faculty Consultants and Student Ambassadors offer insight into how to achieve the results you want with Connect.

# Preface

**W**elcome to your microbiology laboratory course! This laboratory will be unlike any other science laboratory you have taken. You will work with living organisms on a daily basis and study their growth, metabolic capabilities, and a variety of techniques related to the diagnosis and treatment of disease. Microbiology laboratory is a dynamic, hands-on experience with applications that extend into your daily life, from the safety of the food you eat to the maintenance of human health.

In the mid-1990s, we were using a commercially available laboratory manual for our microbiology laboratory course. Unfortunately, we found that we had to constantly amend the manual's content and procedures to fit our course. We developed a handout for the faculty teaching microbiology laboratory classes at Broward College in order to share tips for doing the lab and changes in procedure. After a few years, we realized it would be useful for students to also have this information, and a lab handout was born. Over time, more homegrown exercises were included in the handout. Additionally, extensive online resources were developed to complement laboratory exercises. Then in 2008, we compiled much of this work to create a custom lab manual for our students. We continued to refine the manual through three more editions. Every semester we discussed with colleagues and students what worked well, what did not, and how to make things better. No professor likes spending the second lab period explaining why the exercise students did in the first lab period didn't work. If we could modify the procedure to make it work better, we did. If we could not make it work, we discontinued the exercise.

This is the third edition of our nationally distributed manual. We are grateful to our adopters and their students for trusting us with their microbiology laboratory education and very much appreciate the comments we received from reviewers and colleagues, both positive and negative. This edition incorporates many of your suggestions, including new and modified case files, additional exercises, expanded lab reports, higher-quality photographs, and an increased clinical focus. Thank you for helping us create a better teaching instrument. We are humbled by your support and delighted to play a role in educating your students.

## Our Audience

We understand the level of knowledge students have when they enroll in this course and the level of knowledge necessary to be successful in their chosen fields of study. Between the two of us, we have decades of experience teaching microbiology and microbiology laboratory at all levels of college instruction. This laboratory manual is the compilation of that knowledge and experience. We hope our approach will add to your own cadre of techniques for effective teaching, and if you are new to the field, we hope we can help you and your students avoid many of the pitfalls we encountered. The manual originated as handouts that accompanied other published material and evolved over time to include case files, tips for success, online and electronic supplements, and photographs. This version with its visual, informative, and conversational approach will entice interest and facilitate learning and success in the course.

## Our Organization

The content in this laboratory manual is divided into subsections that address most of the traditional topics covered in an introductory microbiology course, including Basic Techniques (microscopy, aseptic technique, and streak plates), Examination of Bacteria and Eukaryotic Microbes, Staining Techniques, Media, Bacterial Growth, Control of Microbial Growth, Hydrolytic Enzymes, Biochemical Testing, and Unknown Identification. Additionally, other topics examine current trends and applications in microbiology such as Diagnostic and Applied Genetics, Diagnostic Immunology, and Microbes and Infectious Diseases.

## Our Practical Advantage

The topics of exercises in this manual are typically found in most introductory microbiology manuals. However, the presentation and approach to the content are unique. Each exercise incorporates a brief case study, called "Case File," relevant to the activity performed in the exercise. Most of these are modeled from case studies reported to the Centers for Disease Control and Prevention and published in *Morbidity and Mortality Weekly Report*. The Case Files are short, to the point, and realistic. Case studies exemplify the relevancy of the activity. It is important to remember that the techniques and procedures described in this manual are performed on a daily basis in clinical, environmental, and public health laboratories throughout the world. This is *real* microbiology.

## Our Features

The Case File, which begins each exercise, exemplifies the practical applications of that day's activity. Microbiology is a critical component in monitoring and evaluating the safety of our food, environment, and health.

The Case File is followed by specific **Learning Outcomes** for the exercise. In a succinct, bullet format, the Learning Outcomes are spelled out for you. At the completion of each exercise, you, as a student, should be able to explain, demonstrate, and understand the concepts addressed in the Learning Outcomes. They should help you focus your studies.

The **writing style** found in this manual is conversational and concise. This can clearly be seen in the Background and Procedure sections of the exercises. **Background** information includes fundamental concepts and procedural information in a clear and uncluttered format. The **Procedure** section provides activity instructions and illustrations in an easy-to-follow, step-by-step format.

**Tips for Success** are included to ensure your success during the laboratory period. Novice microbiologists, like yourselves, inadvertently make a number of common mistakes during the course. We have compiled the most common mistakes and most often asked questions, to create the "Tips for Success" section to address these error-prone technique steps. By following these tips, students can avoid many of the common pitfalls encountered in the lab. For example, in one

---

**CASE FILE**

Patient C, a 68-year-old woman, was diagnosed with rheumatoid arthritis (RA) 11 years ago. One year after diagnosis, she began treatment with a nonsteroidal anti-inflammatory drug (NSAID). Three years ago, she retired and started backyard gardening in her Idaho community. She discontinued her treatment at that time. One year ago, her arthritis symptoms worsened. Resumption of NSAID therapy had not helped, but the addition of Humira®, a biological response modifier and disease-modifying anti-rheumatic drug, seemed to reduce her symptoms. The patient was evaluated to determine if more aggressive drug therapy was needed.

On physical examination, her elbows, wrists, and metacarpophalangeal joints showed mild swelling and tenderness. Both knees had effusions (extra fluid) and were painful on flexion. She had no subcutaneous nodules (a common symptom of RA) and no other signs of systemic illness. X rays of the affected joints were ordered. In order to find out if her current treatment was sufficient, a test to detect elevated C-reactive protein (CRP) levels was also ordered to see if significant levels of inflammation were still present in Patient C.

©Science Photo Library/ Alamy Stock Photo RF

---

**LEARNING OUTCOMES**

At the completion of this exercise, students should be able to
- properly prepare a negatively stained slide.
- explain the basis for and advantages of negative staining.

---

 Tips for Success

- Make **short** streaks, so that the color zones from adjacent samples don't merge and make it hard to read your results.
- Incubating plates for more than 24 hours may cause zones around the colonies to become too large and interfere with the results. Removing and/or refrigerating plates may be needed.

**Figure 16.3** Growth on the plate.
©Steven D. Obenauf

**Results and Interpretation**

Read your plates against a dark background (for example, your lab bench-top)–you may need to lift the lid a bit. Gram-positive bacteria will grow poorly on this medium or not at all **(figure 16.4a)**. Gram-negative bacteria that are not coliforms will grow well but are basically colorless **(figure 16.4b)**. Coliform bacteria (lactose fermenters) will grow well and the colonies will be pink with purple in the center **(figure 16.4c)**. Rapid or strong lactose fermenters will grow well and will appear dark purple with a distinct metallic green color **(figure 16.4d)**.

exercise, both a warm water bath and a boiling water bath are used. But they cannot be used interchangeably! So the tip for success is to make sure you place the appropriate tube in the correct water bath.

The number of **photographs and illustrations** in the manual is unprecedented. Every exercise in this manual contains color photographs, including equipment, techniques, and experimental results. These photographs were taken by the authors while performing the exercises in the manual and therefore reflect actual experimental results or observations. Microbiology is a very visual science. This third edition includes more and higher-quality photographs that will guide you, the student, through the exercises and advance student learning. Therefore purchasing a supplemental hard copy atlas will not be necessary.

The **Results and Interpretation** section provides an illustration, explanation, or photograph of the exercise activity. This provides a visual or contextual comparison for the results you obtain and a guide to the meaning of those results.

The **Lab Reports** at the end of each exercise are designed to enhance student learning. Each report includes an area for students to report their experimental results, including the option of taking and attaching a digital photograph rather than drawing. Additionally, the reports include questions that utilize both recall learning and critical thinking skills. In this edition, we have increased the number of questions in some lab reports to broaden the coverage of the laboratory content.

The online **Instructor's Resource Guide,** written by the laboratory manual authors, provides additional teaching tips for instructors. For convenience, the manual also cites sources and recipes for media, samples, and slides.

## Digital Tools for Your Lab Course
## McGraw-Hill Connect®

McGraw-Hill Connect is a digital teaching and learning environment that saves students and instructors time while improving performance over a variety of critical outcomes.

- Instructors have access to a variety of resources, including assignable and gradable interactive questions based on textbook images, case study activities, videos, and more.

- Digital Lecture Capture: Get Connected. Get McGraw-Hill Tegrity®. Capture your lectures for students. Easy access outside of class anytime, anywhere, on just about any device.

## Customize your course materials to your learning outcomes!

### Create what you've only imagined.

Introducing McGraw-Hill Create™–a self-service website that allows you to create custom course materials–by drawing upon McGraw-Hill Education's comprehensive, cross-disciplinary content. Add your own content quickly and easily.

## To the Student

This manual has been written with you in mind. We have used our years of laboratory teaching experience to identify and relate the fundamental concepts that are important to understanding the science and to your success in your future field. We have included case studies to exemplify the relevance of the science. The extensive visual imagery will help you understand and learn the content. The writing style in the manual is less formal and more conversational to increase its readability. We have tried to avoid jargon and a more formal writing style because we want you to feel comfortable with this laboratory manual and use it as a resource during the course of the laboratory period and beyond.

Most importantly, we hope you will experience the thrill of bench science and share some of the enthusiasm we have for microbiology, a field of science that is dynamic, exciting, and touches every aspect of your life.

Steve Obenauf

Susan Finazzo

# Changes to the Third Edition

## Tools of the Laboratory: Basic Techniques

### Exercise 1 Use of the Microscope
- Minor changes to background and procedure sections

### Exercise 3 Isolation Streak Plate
- Option of using blood agar plates added

## Eukaryotic Cells and Microorganisms

### Exercise 4 Microbial Phototrophs: Algae and Cyanobacteria
- Modifications to the background material to include more information on the health impacts of cyanobacterial or algal toxins

### Exercise 6 Fungi
- Modifications to the background material to expand discussion of opportunistic and pathogenic fungi

### Exercise 7 Bacterial Motility: Hanging Drop Method
- Modifications to the background material

## Methods for Microscopic Analysis: Staining

### Exercise 9 Negative Stain
- Modifications to the case file and results and interpretation section

### Exercise 12 Acid-Fast Stain (Ziehl-Neelsen and Kinyoun Methods)
- New image of stained cells

### Exercise 13 Endospore Stain (Schaeffer-Fulton Method)
- Modifications made to the case file
- Modifications made to the questions in the lab report

## Methods for Culturing and Diagnosing Microbes: Media

### Exercise 14 Blood Agar
- Modifications made to the case file
- Linking of this exercise to the new rapid immunoassays exercise

### Exercise 15 Mannitol Salt Agar
- Modifications to the background information

### Exercise 16 Eosin Methylene Blue Agar

- Replacement of case file with a new one involving homeowners' concerned about the safety of their water supply
- Option added for a more inquiry-based approach in which students can bring their own water sample
- Added image of an uninoculated plate

## Microbial Nutrition, Ecology, and Growth

### Exercise 18 Chemical Content of Media

- Added information to the background about the term *fastideous* relative to the growth requirements of disease-causing bacteria

### Exercise 19 Osmotic Pressure and Growth

- Replacement of case file with a new one involving a *Vibrio vulnificus* infection

### Exercise 20 Oxygen and Growth: Aerotolerance

- Modifications made to the case file
- Modifications made to the organisms used
- Unclear figure removed

### Exercise 21 Oxygen and Growth: Catalase

- Changes made to the case file

### Exercise 22 Oxidase

- Replacement of case file with a new one involving a *Pseudomonas* infection in a burn patient
- Linking of this exercise to the antimicrobial susceptibility exercise

## Food and Water Safety

### Exercise 23 Plate Count

- Modification of the title and background to reflect current methods and use of the term *Heterotrophic Plate Count (HPC)*
- New image–reading volume of micropipette

## Treatment and Control of Microbial Growth

### Exercise 24 Killing by Ultraviolet Light

- Modification of the organisms used to provide clearer results

### Exercise 25 Antimicrobial Susceptibility Testing (Kirby-Bauer Method)

- Significant changes to antimicrobials used to reflect those most commonly used in medicine today
- Addition of a narrow-spectrum drug with activity against gram-negative bacteria
- Modification to the case file
- Linking of case file and activities to other exercises
- Other modifications

# Microbial Genetics and Genetic Engineering

## Exercise 26 ELISA and Using Nucleotide BLAST (Dry Lab)

- New addition to the lab manual
- Case file of a child with persistent measles virus infection leading to SSPE
- Use of ELISA for initial diagnosis and gene sequencing to determine the strain and genotype
- Use of the BLAST tool from the National Center for Biotechnology Information to identify the virus isolate

## Exercise 27 Analysis of DNA: Electrophoresis

- Changes made to background information

## Exercise 28 Gene Transfer: Transformation

- New option added to the procedure
- New image

# Infectious Diseases

## Exercise 29 Parasitic Protozoa

- Modifications to the case file and background information
- Improved images of the *Giardia* cyst in photomicrograph 29.2 and the *Plasmodium* in 29.4

## Exercise 30 Parasitic Worms

- Modification of the case file
- Improved images of the hookworm larva and pinworm egg photomicrographs in figure 30.3 and of the *Trichinella* cyst in figure 30.5

## Exercise 31 Transmission: Vectors of Disease

- Modification of exercise title for further clarity
- Replacement of case file with a new case with information and data about *Aedes* traps and Chikungunya virus transmission
- Addition of Zika to the list of mosquito-borne diseases
- New photo to show the nature of vector slides and how to view them properly
- Question added to lab report involving analysis of a data table in the new case file

# Diagnosing Infections

## Exercise 32 Gel Immunoprecipitation (Immunodiffusion)

- Information on vaccines added to the background

## Exercise 33 Passive (Indirect) Agglutination

- Additional information on the two types of arthritis

## Exercise 34 Rapid Immunoassays

- New addition to the manual, using a rapid strep test as an example of rapid immunoassays used in medical diagnosis
- Linking of this exercise to the blood agar and antimicrobials labs

## Exercise 35 Hemagglutintin Inhibition Assay

- New addition to the manual
- Case file and background dealing with influenza virus and antigenic variation
- Use of the assay to identify the infecting strain of virus

## Exercise 36 EnteroPluri-Test

- New language about variability in codes
- Three photographs replaced with new images

# Environmental Microbiology

## Exercise 37 Soil Microbiology

- Changes made to the procedure section

# Hydrolytic Enzymes and Disease Mechanisms

## Exercise 38 DNase Test

- Modification of case file to discuss the role health care workers can play in preventing central line infections

## Exercise 39 Gelatin Hydrolysis

- "Mini unknown" question added to the lab report to give students practice for the critical thinking needed for exercise 49 (unknown identification)
- Language about the role of protease in infectious disease added, along with an out-of-classroom activity

## Exercise 40 Starch Hydrolysis

- Replacement of case file with one involving ventilator-associated pneumonia in a neonatal intensive care unit

# Biochemical Testing and Diagnosis of Disease

## Exercise 42 Phenol Red Broth

- Language added to the case file to better link it to the tests being done in the exercise
- Restructured table for recording results

## Exercise 43 Triple-Sugar Iron Agar

- "Mini unknown" question added to the lab report to give students practice for the critical thinking needed for exercise

## Exercise 44 Litmus Milk

- Modifications made to the case study

## Exercise 45 IMViC (Indole, Methyl Red, Voges-Proskauer, Citrate) Reactions

- Modification of MRVP procedure
- Modification of lab report to reflect student use of photography with mobile devices

## Exercise 46 Urease Test

- "Mini unknown" question added to the lab report to give students practice for the critical thinking needed for exercise 49 (unknown identification)

## Exercise 47 Nitrate Reduction

- Modifications to the case file and background

# Putting It All Together

## Exercise 48 "Known" Bacteria for Biochemical Testing

- Additional organisms listed—*Bacillus megaterium, Campylobacter jejuni,* and *Corynebacterium xerosis*
- Modification to the procedure

## Exercise 49 Identification of Unknown Bacteria

- Organisms added to the options—*Bacillus megaterium* and *Corynebacterium xerosis*
- Additional detail in the nitrate results
- More flexibility for instructors in the procedure section
- References to prior exercises added to the procedure section

## Hand Washing

- This exercise can be accessed online

# Acknowledgments

As any professor can tell you, without pure cultures, properly made media, and adequate supplies, teaching laboratory is pretty much impossible. Lab technicians who made this manual possible and gave suggestions for improvement include Yury Mazo, who has assisted ably for many years, first as a lab technician and later as laboratory manager; Rani Aranda; Ryan Ebanks; and Todd Nims. Our colleague Dr. Derek Weber was involved in the first three custom editions of the manual, and we value his contributions. Lisa Burgess, Jessica Digirolamo, and Idelisa Ayala have our sincere thanks for their thorough review of the initial editions of the manual.

Thanks to Kathy Loewenberg, Michelle Watnick, and Nancy Myers for their initial interest and support. Our Product Developer, Darlene Schueller, turned text and images into structured labs with such speed we worried about caffeine overdose. We are also thankful for Sr. Portfolio Manager Marija Magner, Marketing Manager Valerie Kramer, Content Project Manager Mary Jane Lampe, Assessment Content Project Manager Christina Nelson, Sr. Project Manager Sarita Yadav, Content Licensing Specialist Lori Hancock, and Designer Matt Backhaus for their assistance and support.

Thanks to our students. We have been inspired by so many of them over the years. The questions they asked and the way they approached (or avoided) their studies have shaped our teaching. We learned from their mistakes as well as our own as we developed this manual.

We are grateful to Lisa Burgess for her review of several exercises.

And a special thanks to our family members–Gail, Scott, Lauren, Brian, Jonathan, Adam, Emerson, Kira, and Grayson–for their unfailing support and the joy they bring to our lives.

# Inside the Lab Exercise

Methods for Microscopic Analysis: Staining

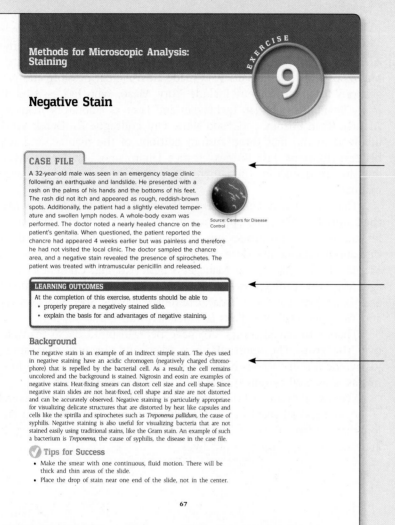

**EXERCISE 9**

## Negative Stain

### CASE FILE

A 32-year-old male was seen in an emergency triage clinic following an earthquake and landslide. He presented with a rash on the palms of his hands and the bottoms of his feet. The rash did not itch and appeared as rough, reddish-brown spots. Additionally, the patient had a slightly elevated temperature and swollen lymph nodes. A whole-body exam was performed. The doctor noted a nearly healed chancre on the patient's genitalia. When questioned, the patient reported the chancre had appeared 4 weeks earlier but was painless and therefore he had not visited the local clinic. The doctor sampled the chancre area, and a negative stain revealed the presence of spirochetes. The patient was treated with intramuscular penicillin and released.

Source: Centers for Disease Control

### LEARNING OUTCOMES

At the completion of this exercise, students should be able to
- properly prepare a negatively stained slide.
- explain the basis for and advantages of negative staining.

### Background

The negative stain is an example of an indirect simple stain. The dyes used in negative staining have an acidic chromogen (negatively charged chromophore) that is repelled by the bacterial cell. As a result, the cell remains uncolored and the background is stained. Nigrosin and eosin are examples of negative stains. Heat-fixing smears can distort cell size and cell shape. Since negative stain slides are not heat-fixed, cell shape and size are not distorted and can be accurately observed. Negative staining is particularly appropriate for visualizing delicate structures that are distorted by heat like capsules and cells like the spirilla and spirochetes such as *Treponema pallidum*, the cause of syphilis. Negative staining is also useful for visualizing bacteria that are not stained easily using traditional stains, like the Gram stain. An example of such a bacterium is *Treponema*, the cause of syphilis, the disease in the case file.

### ✔ Tips for Success

- Make the smear with one continuous, fluid motion. There will be thick and thin areas of the slide.
- Place the drop of stain near one end of the slide, not in the center.

67

**Case file.** Case files derived from real-life scenarios are used to exemplify the depth and breadth of the field of microbiology.

**Learning outcomes.** Specific learning outcomes are provided for each exercise to highlight core concepts and focus student learning.

**Background.** Background information is provided for each exercise. It may include historical context, foundational knowledge, or current applications.

**Illustrations and Photos.** Full-color photographs and illustrations visually explain the course content.

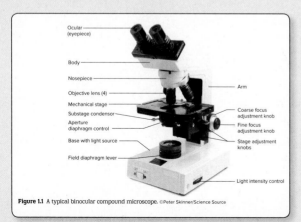

Ocular (eyepiece)
Body
Nosepiece
Objective lens (4)
Mechanical stage
Substage condensor
Aperture diaphragm control
Base with light source
Field diaphragm lever
Arm
Coarse focus adjustment knob
Fine focus adjustment knob
Stage adjustment knobs
Light intensity control

**Figure 1.1** A typical binocular compound microscope. ©Peter Skinner/Science Source

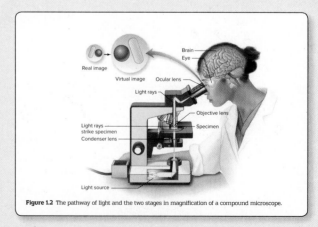

Real image
Virtual image
Brain
Eye
Ocular lens
Light rays
Objective lens
Light rays strike specimen
Condenser lens
Specimen
Light source

**Figure 1.2** The pathway of light and the two stages in magnification of a compound microscope.

# Lab Report 25

## Antimicrobial Susceptibility Testing (Kirby-Bauer Method)

### Your Results and Observations

Record your results in the table. If instructed, you may photograph your plates with your mobile device.

| Antibiotic | Code | Patient B Zone (mm) | Result S/I/R* | Patient S Zone (mm) | Result S/I/R | Patient X Zone (mm) | Result S/I/R |
|---|---|---|---|---|---|---|---|
| Azithromycin | AZM-15 | | | | | | |
| Ciprofloxacin | CIP-5 | | | | | | |
| Penicillin | P-10 | | | | | | |
| Polymyxin B | PB-300 | | | | | | |
| Tetracycline | TE-30 | | | | | | |
| Trimethoprim | TMP-5 | | | | | | |
| | | | | | | | |
| | | | | | | | |

*S = sensitive, I = intermediate, R = resistant.

### Interpretation and Questions

1. Based on your results, which patient's infection appears to b

_____  
_____  
_____  
_____

173

---

# Lab Report 45

## IMViC (Indole, Methyl Red, Voges-Proskauer, and Citrate) Reactions

### Your Results and Observations

Record your results after incubation in the table. You may also wish to use your mobile device to (carefully) photograph your various tubes.

| Organism | SIM Sulfide (+ or -) | SIM Indole (+ or -) | SIM Motility (+ or -) | MRVP MR (+ or -) | MRVP VP (+ or -) | Citrate Slant color | Citrate Citrate (+ or -) |
|---|---|---|---|---|---|---|---|
| *Salmonella typhimurium* | | | | | | | |
| *Escherichia coli* | | | | | | | |
| *Enterobacter aerogenes* | | | | | | | |
| *Klebsiella pneumoniae* | | | | | | | |

### Interpretation and Questions

1. The IMViC tests are based on some enzymatic/physiological process. Name the substrates for the enzymes in the indole, MRVP reactions, and citrate test.

_____  
_____  
_____  
_____

2. Is it possible for an organism to be MR⁺ and VP⁺? Why or why not?

_____  
_____  
_____  
_____

313

---

**Lab reports.** Lab reports include areas to record experimental results and observations. Additional questions are included that require both recall learning and critical thinking.

# Introduction

To most students, at least in the beginning, microbiology is "just some class I have to take." In fact, whether we want to know it or not, microbiology is a major part of the world in which we live. Not only are microbes and the immune system involved in a wide variety of diseases but they also play a positive role in the environment, diagnostic medicine, and the production of food and pharmaceuticals. They are involved in everything from the coffee you drink in the morning to the cold you always seem to catch just before taking exams. To show how these labs fit into the world around us, we have used a scenario and case file approach in many of the labs in this manual and tried to emphasize critical thinking, interpretation, and understanding of results. In addition, a number of labs are connected to each other by examining different parts of the same situation (such as the antimicrobial sensitivity testing and DNA finger-printing labs and the wound infection scenario). Most lab days, you will probably be doing more than one exercise. Be patient: As the semester goes on, your skills and understanding of what you are doing will continue to improve.

## Some Situations and Case Files You'll Be Working With

**Exercise 10:** In the capsule stain exercise, a first responder acquires meningitis from a patient who has to be transported to the emergency room.

**Exercise 11:** In the Gram-stain exercise, the patient has a corneal infection.

**Exercise 13:** The case file in the spore stain is a baby who had been fed unpasteurized honey. You'll be doing the stain that is part of the testing needed to see if it is the cause of the baby's muscle weakness.

**Exercises 14, 25, and 34: Patient A** has a sore throat. What are the cause and best treatment?

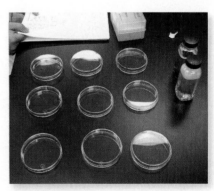
©Steven D. Obenauf

**Exercise 15:** In the **Local Festival scenario,** you'll be looking at Sample M for the cause of an outbreak of food-borne illness. Many other case files also deal with diseases associated with food. You'll be performing procedures associated with investigating symptoms associated with oysters, fish, ham, lobster rolls, fried rice, and tomatoes.

**Exercise 16:** Two groups of homeowners are concerned about water safety.

**Exercise 19:** In the **Dairy scenario,** we'll be testing milk to see if it is safe to drink. You'll also be looking at water from the Chattahoochee River and a swimming pool.

**Exercise 22: Patient G** became infected after a kidney transplant. Which gram-positive bacterium is causing it?

**Exercises 23 and 25: Patient B** is a burn patient with an infection. You'll be testing to find the best antibiotic to treat him or her with.

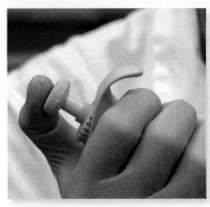
©Steven D. Obenauf

©Steven D. Obenauf

©Steven D. Obenauf

**Exercises 25, 27, 34: Patient X** has a wound infection. We'll be using several different tests (the antimicrobial susceptibility test, DNA fingerprinting, and EnteroPluri-Test labs) to find the causative organism, the source, and the best antibiotic to treat the patient with.

**Exercise 30:** In the **Day Camp scenario,** we'll be looking for why some children have itchy skin lesions.

**Exercise 31:** Testing is carried out to try and prevent transmission of Chikungunya and Zika.

**Exercise 32: Patient T** needs to find out whether or not to get a vaccine booster.

**Exercise 33: Patient C** has rheumatoid arthritis. You'll be finding out if her treatment is still working.

**Exercise 43:** In the TSI agar lab, you'll see a patient with an oral infection following a root canal.

**Exercise 46:** A patient has burning pain in her upper abdomen. A noninvasive test used to detect the bacteria associated with peptic ulcers looks for a particular enzyme. You'll do a test looking for the same enzyme.

## Tips for Success in Lab

- **Before lab.** What you do before lab can often be even more important than what you do during lab. Reading the labs and doing any Web-based or hard-copy pre-lab activities your professor assigns will help you use your time more effectively, reduce the chance for mistakes, and make the time in lab a more valuable learning experience.

- **During lab.** Attentiveness and participation are key. Even for exercises done in teams, be as involved as possible. You will learn more by doing than by watching. Pay attention at every stage of the exercise.

- **After lab.** Analyze the results of each exercise and see how they relate to the associated case file.

# Laboratory Safety, Equipment, and Materials

## CASE FILE

Your patient, a 24-year-old student, has a temperature of 39.1°C (102.4°F) and an abscess on her right index finger. She reports to you that 2 days ago, while in the microbiology lab, she tipped a broth tube of *Staphylococcus aureus* to look at it and the broth spilled all over her hand. She had a cut on her finger at the time and did not wear gloves or cover it in any way. She was unsure if she washed or simply rinsed her hands after the spill and how long she waited to do that.

Source: CDC/Janice Haney Carr

## SAFETY: Some Initial Thoughts

1. The microbiology lab presents more potential risks than virtually any other lab you will ever take. Paying attention to safety is particularly important here.

2. You will be working with fire, bacteria, and toxic chemicals this term. The following rules are intended to keep you and everyone else in the lab safe.

(a)

(b)

©Steven D. Obenauf
**Figure 1**

## SAFETY: Your Work Area

1. At the beginning and end of each lab, clean your work area with the **spray disinfectant (figure 1)** provided.

2. Keep your work area uncluttered–bring only what you need: your lab manual and pens or pencils. Leave all other books, backpacks, purses, and so on in the area of the lab where biohazardous materials are not handled.

3. If you are using one, the Bunsen burner **(figure 2)** may well be the most dangerous thing in the lab. When using one, try and keep it near the center of your bench. Be careful when handling it and working around it. **Make sure the gas jet (figure 3) is turned off before you leave the laboratory.**

©Steven D. Obenauf
**Figure 2**

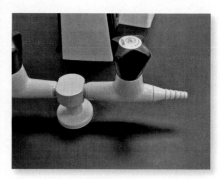

©Steven D. Obenauf
**Figure 3**

**Figure 4**

## SAFETY: Personal Safety

1. **Eating, drinking, and chewing gum are not allowed in the laboratory.**
2. You should **wash your hands** after, and even during, lab with antibacterial soap **(figure 4).**
3. **Lab coats** should be worn at all times. This will protect you against not only microbes but also chemicals and your Bunsen burner flame.
4. Gloves will be available in the lab and may be needed for some labs to protect you against microbes or stains. Be careful around flames when wearing latex gloves.
5. Eye protection should be worn and is particularly important for labs dealing with certain chemicals or ultraviolet (UV) light.
6. In order to protect your feet against dropped material, sandals or open-toed shoes must not be worn in the laboratory.
7. Students with long hair should tie it back to avoid accidents.
8. If you have a cut or an injury before coming to the lab, it should be covered with Band-Aids and possibly gloves while working in the lab.
9. If you are using a **smartphone** or other portable device to take pictures in the laboratory, or a tablet or computer to view online materials, handle it with care to avoid contaminating it with bacteria or chemicals.

## SAFETY: Disposal of Materials

1. Petri dishes and many other biohazardous items should be discarded in a large orange, autoclavable, biohazard bag **(figure 5)** or biohazard bin. Please be aware that disposing of sharp objects such as pipettes or toothpicks in this bag could get somebody hurt.
2. When disposing of test tubes, place them in the appropriate rack or bin in the disposal area. Remove the caps and place them in the provided basket or container.
3. Broken glass, pipette tips, toothpicks, and other small, sharp objects should be placed only in a **sharps container (figure 6).**
4. Used pipettes should be placed in a discard bucket or in a large sharps container.

**Figure 5**

**Figure 6**

5. You won't be using the regular garbage can much in this lab. Do not throw biohazardous material of any kind in the regular garbage–ever.

6. Do not remove materials or cultures from the lab.

## SAFETY: Dealing with Accidents

1. Know the location of the fire extinguishers and fire blankets **(figure 7).**

2. Know the location of the eyewash station **(figure 8)** and safety shower.

3. If you injure yourself in any way during the laboratory, notify your professor or the lab staff *immediately.*

4. If you smell gas, notify your professor or the lab staff *immediately.*

5. If you spill bacteria or other biohazardous material on a surface, cover it with a paper towel and soak the paper towel with disinfectant.

6. Do not try to pick up broken glass.

7. If you spill bacteria or other biohazardous materials on yourself, wash the exposed area with antibacterial soap and take any other needed measures (such as the use of the eyewash or safety shower). **A common cause of spills is tipping broth tubes sideways or picking them up by the cap instead of holding the tube itself.**

## SAFETY: Some Final Thoughts

1. Safety doesn't have to be complicated: Be observant and use common sense!

2. Read the procedures and understand what you are doing **BEFORE** coming to lab. Your pre-lab study will make you aware of any potential dangers in advance and reduce the likelihood of making mistakes.

3. Sign and return the safety contract you are given.

## Caring for Microscopes

1. Always carry your microscope with two hands.

2. The microscopes and storage shelves may be **numbered.** If they are, at the end of the lab, return your microscope to the shelf with the correct number.

3. Use immersion oil **only** with your oil immersion (100×) objective. If you get oil on your high dry (40×) objective, you (and everyone who uses the microscope after you) will no longer be able to see properly through it.

4. At the end of the lab period, students should **clean all immersion oil off the objective lens** before putting the microscope away to avoid damaging it. Clean all optical components (objective, ocular, condenser, and illuminator) with lens paper **(figure 9)** and lens cleaner. DO NOT use paper towels–they will scratch the glass. Also, clean off any oil that may have gotten on the microscope stage.

5. When putting your microscope away, always put the scanning (4×) objective **(figure 10)** in position to avoid possible damage to the longer, higher-power objectives.

## Microscope Slides

1. Students wishing to keep any of their stained slides may place them in slide boxes for future reference and put them in the slide drawer for their team or class. Label the box lid with masking tape.

©Steven D. Obenauf
**Figure 7**

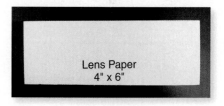
©Steven D. Obenauf
**Figure 8**

©Steven D. Obenauf
**Figure 9**

©Steven D. Obenauf
**Figure 10**

©Steven D. Obenauf

**Figure 11**

©Steven D. Obenauf

**Figure 12**

2. Lens paper, cleaner, immersion oil, and bibulous paper are community property for everyone to use. Return them to where you got them.

3. If you break a slide, throw it away in the **sharps container,** not the regular garbage.

## Bunsen Burners

1. Bunsen burners should be put away at the end of the lab. Make sure they have cooled off before handling them!

2. Place burners near the **center of the lab bench**–not near the edge.

## Handling Media and Cultures

1. ALL stored media (in incubators, room temperature, or refrigerators) **MUST BE PROPERLY LABELED** with the student's or team's initials, lab day and time, and date.

2. When labeling **plates,** use a permanent marker to write on the **bottom** of the plate, not the lid. Keep your writing small and around the outside edge, so that it won't get in the way of reading your results **(figure 11).**

3. Always place plates in the incubator **lid side down** to prevent water from the lid dropping onto your cultures.

4. When labeling **tubes,** use a permanent marker to write **on the glass. DO NOT** mark on the caps or use masking tape to label the tubes **(figure 12).**

5. Do not store inoculated media in the student drawers.

6. Discard used media as soon as you have read the results, so that the incubators do not become overcrowded.

7. All "stock" bacterial cultures should be discarded **after each lab** by the students unless otherwise instructed. Tubes should be discarded in the biohazard bins in the disposal area.

## List of Bacteria and Fungi to Be Used in the Labs

You will be working with a variety of bacteria in this laboratory throughout the semester. In some cases, the plate or tube that they are in may be labeled with the name of the organism. In others, you may use the color of the cap or a number on the tube to let you know what is in it. A list of the numbers that will be used follows. Those bacteria with the lowest numbers are the ones used most often.

Many of these bacteria are relatively harmless ones that are normal biota or from the environment. Others are potentially harmful and pose significant risks. You should handle **all** of them with attention and respect.

| Number | Organism | Number | Organism |
|---|---|---|---|
| 1 | *Escherichia coli* | 14 | *Mycobacterium smegmatis* |
| 2 | *Bacillus cereus* | 15 | *Campylobacter jejuni* |
| 3 | *Enterobacter aerogenes* | 16 | *Proteus vulgaris* |
| 4 | *Micrococcus luteus* | 17 | *Pseudomonas aeruginosa* |
| 5 | *Serratia marcescens* | 18 | *Rhizopus* spp. |
| 6 | *Staphylococcus aureus* | 19 | *Saccharomyces cerevisiae* |
| 7 | *Alcaligenes faecalis* | 20 | *Salmonella typhimurium* |
| 8 | *Bacillus megaterium* | 21 | *Shigella dysenteriae* |
| 9 | *Bacillus subtilis* | 22 | *Staphylococcus epidermidis* |
| 10 | *Clostridium sporogenes* | 23 | *Streptococcus mitis* |
| 11 | *Corynebacterium xerosis* | 24 | *Streptococcus pyogenes* |
| 12 | *Enterococcus faecalis* | 25 | *Streptococcus sanguinis* |
| 13 | *Klebsiella pneumoniae* | | |

# Use of the Microscope

## CASE FILE

The public health department was contacted in the early spring by a local homeowners association concerned about the foul odor, slimy consistency, and murky appearance of water in an artificial pond in their community. The 40-acre pond is spring fed and has no natural tributaries. Thirty homes in the development and a community park share the pond shoreline. The pond is used for recreational purposes such as swimming and fishing. Water from the pond is not potable or used for human consumption. Residents are worried that the pond water may be a threat to the health of the community members.

©McGraw-Hill Education/
Pat Watson, photographer

This scenario could result from pond turnover in the spring, over-fertilization of the landscape by residents, or some other unusual cause. Your job as the staff biologist for the public health department is to sample and microscopically examine the pond water to determine if there is a biological agent responsible for these "symptoms" and, if so, to make recommendations to the homeowners association.

### LEARNING OUTCOMES

At the completion of this exercise, students should be able to

- identify the parts of a microscope and explain their function.
- define terms relating to microscope functionality such as *resolution*, *contrast*, *magnification*, *parfocal lens*, and *field of view*.
- demonstrate how to prepare a wet mount and how to focus a microscope.
- locate and focus a specimen using oil immersion techniques.

## Background

A microscope is an instrument for magnifying and looking at small things. The first use of a glass lens (a simple microscope) as an aid to vision was described around the year AD 1270. Later, around AD 1600 in Holland, two lenses were put together at opposite ends of a tube to produce the first compound microscope. Today, microscopes are an essential tool in nearly every biological sciences laboratory. In microbiology laboratory classes, the microscope is used in nearly every laboratory period. It is **essential** that you develop good microscopy skills.

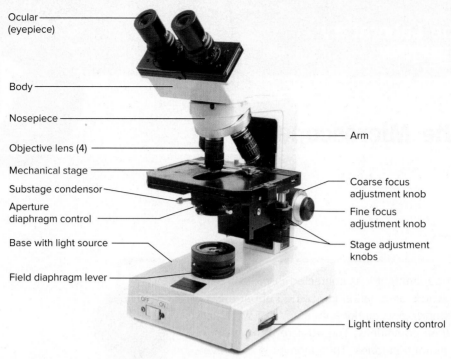

Ocular (eyepiece)

Body

Nosepiece

Objective lens (4)

Mechanical stage

Substage condensor

Aperture diaphragm control

Base with light source

Field diaphragm lever

Arm

Coarse focus adjustment knob

Fine focus adjustment knob

Stage adjustment knobs

Light intensity control

**Figure 1.1** A typical binocular compound microscope. ©Peter Skinner/Science Source

The microscope you will use this semester is a binocular compound light microscope **(figure 1.1).** Binocular means the microscope has two oculars or two eyepieces for viewing the specimen. A compound microscope has at least two lenses or mirrors used to produce an image. Visible light is the source of illumination.

The compound microscope is designed for the examination of transparent specimens–that is, light must be able to pass through them. Specimens are mounted on a glass slide covered by a very thin glass or plastic coverslip. The slide rests on the microscope's stage and is held in place by the mechanical stage. The image is produced when light from an illuminator below the stage passes through the specimen, then through the body tube of the microscope, and then through the ocular to the eye. At the lower end of the body tube, objective lenses screw into the revolving nosepiece. The shortest of these, the **scanning objective** (4×), is used to quickly locate your specimen by viewing a large area of the slide. As magnification increases, smaller portions of the specimen are viewed through the microscope. The medium power objective (10×) is the next longest objective; the **high (power) dry objective** (40×) is the next longest objective; and the longest objective is the 100×, or **oil immersion objective.** The ocular (10×) lens further magnifies the image seen by the student.

## Microscope Parts and Their Function

**Ocular/eyepiece.** A binocular microscope has two oculars or eyepieces. The oculars magnify the image 10-fold. The oculars move to accommodate your interpupillary distance–that is, the distance between the pupils of your eyes. To adjust the oculars, grab the oculars, move them closer together or farther apart while looking into them. The oculars are correctly positioned when you see a single circle of light. If you see two circles of light, continue moving the oculars until the circles merge. Keep both eyes open while viewing your slides. Closing one eye tightens facial muscles and after an extended

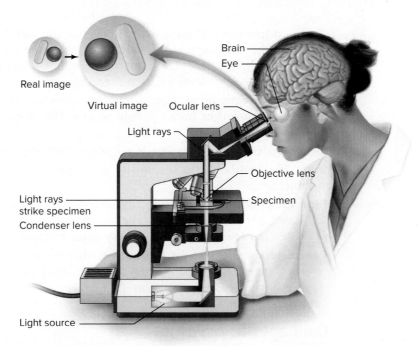

**Figure 1.2** The pathway of light and the two stages in magnification of a compound microscope.

period of time can lead to headaches and decrease your enjoyment of using the microscope.

**Body.** The body contains mirrors that bend the light into the ocular. The orientation of the microscope's optical elements causes the image to be reversed (right to left) and inverted (top to bottom), producing a virtual image of the specimen **(figure 1.2).**

**Nosepiece.** The objectives are attached to the nosepiece. Grab the black ring of the nosepiece and rotate the objectives. You should hear a clicking sound when the objectives are correctly aligned.

**Objectives.** The objectives contain lenses that magnify the image. Your microscope has 4×, 10×, 40×, and 100× objectives. Magnification and resolution are more thoroughly discussed later.

**Arm.** The arm supports the nosepiece and holds the on/off toggle and focus adjustment knobs.

**Stage/mechanical stage.** The stage is the black, flat platform on which your slide rests. The mechanical stage is the silver or black apparatus that actually holds the slide and moves the slide to position it on the stage.

**Stage adjustment knobs.** The stage adjustment knobs move the mechanical stage. The mechanical stage control consists of two knobs. One knob moves the mechanical stage from right to left. The other knob moves the mechanical stage from front to back.

**Light intensity control.** This knob increases or decreases light output from the light source. At lower magnifications and with unstained specimens, you may need to decrease the light output with this knob. When using the oil immersion lens with stained slides, you should maximize light output.

**Substage condenser.** The condenser is made of at least two lenses that focus the light passing through the specimen and improve image sharpness. Although the condenser can be moved up and down, for most applications it should be in its uppermost position, close to the slide.

**Aperture diaphragm control.** The diaphragm controls the amount of light entering the substage condenser from the light source. This is probably your most important mechanism for adjusting light. Set

9

the diaphragm half open when using the 10× and 40× objectives. Open the diaphragm completely when using the oil immersion lens. Too much light can bleach out your specimen. You need to use the appropriate light level for each magnification.

**Field diaphragm lever.** This diaphragm opens and closes to increase and decrease light from the light source. This diaphragm should remain open.

**Light source.** The illuminator is a halogen bulb that produces light.

**Base.** The base holds the illuminator and supports the rest of the microscope.

**Fine focus adjustment knob.** By rotating the fine focus adjustment knob, it is possible to bring the image into sharp focus. This knob raises and lowers the stage in very small increments.

**Coarse focus adjustment knob.** The large coarse focus adjustment knob moves the stage up and down by larger increments and allows you to bring the image into focus.

## Concepts and Terms

A microscope has three distinct properties related to its function: **magnification** (the number of times the image is enlarged), contrast (striking difference between similar objects), and **resolution** (the ability of a lens system to show small, close together objects as being separate). The unaided human eye can normally distinguish objects that are no smaller than 0.1 millimeter (mm) (100 micrometers [µm]). The light microscope cannot resolve objects smaller than 0.25 µm. The light microscope then enlarges the resolved image to dimensions that can be perceived by the human eye. Magnification is limited by resolution. The limit of useful magnification by the light microscope is about 1,500×. Magnification beyond this point would result in useless, blurred images.

Resolution of a microscope is limited by the wavelength of the illumination source and dependent upon the physical characteristics (numerical aperture) of the lenses. When using visible light to illuminate your specimen, objects smaller than half the wavelength of light are not visible to the observer. Since the wavelength of visible light is 0.5 µm, a light microscope cannot resolve objects smaller than 0.25 µm. Bacteria that will be observed in this lab are typically less than 4 µm in size.

Most bacteria in their natural state are uncolored. Their clear, watery cytoplasm, contained within their uncolored cytoplasmic membrane, and cell wall make them challenging to view with the microscope. Increasing the contrast between the cells and the background makes it easier to see the cells. **Contrast** is a measure of the differences in appearance between two objects. Staining is one way to increase contrast. Staining cells imparts color to the cells, which increases the contrast with the bright background of the light microscope. **Adjusting the light** passing through the specimen is another way to adjust contrast. This very simple method is one many students fail to utilize. As a general rule, the lower the magnification, the lower the level of light needed for good contrast. The higher the magnification, the higher the level of light needed for good contrast.

*Magnification* refers to how many times larger the image appears compared to the actual size of the specimen. The total magnification of an image is determined by multiplying the magnification of the ocular by that of the objective lens–for example, to determine the total magnification when using the scanning objective, you would multiply the power of the ocular (10×) by the objective power of the scanning objective (4×) to get 40× as the total magnification **(table 1.1).**

### Table 1.1 Total Magnification

| Objective | Ocular magnification | Total magnification |
|---|---|---|
| 4× (scanning) | 10× | |
| 10× | 10× | |
| 40× (high dry) | 10× | |
| 100× (oil immersion) | 10× | |

**Resolution.** The ability of a lens system to distinguish as separate entities objects that are very small and very close together. Resolution is determined by the quality of the lens and the wavelength of light used.

**Magnification.** Making a small object appear larger.

**Parfocal lens.** Parfocal lenses maintain focus of a specimen as objective lenses are switched from one objective power to another. Once the specimen is focused, the objectives can be changed and the specimen will still be in relatively good focus at the next higher power. Only fine adjustment should be required from this point.

**Depth of field.** Because all specimens have thickness, it is possible to focus at different planes in the specimen. The thickness of this plane of focus is known as the depth of field. Depth of field is particularly noticeable in thick specimens. Focusing up and down through the specimen may be necessary to observe structures of interest. The depth of field decreases as magnification increases.

**Field of view.** When looking into the oculars, the portion of the slide that is visible in the circle of light is called the field of view. The actual size of the field of view decreases as magnification increases. It is possible to measure the actual size of the field of view and hence the size of the specimen.

**Working distance.** The distance between the tip of the objective and the stage.

## Oil Immersion Microscopy

Light bends or refracts when it moves from one medium to another. So as light from the light source passes into the slide, it refracts **(figure 1.3).** It bends or refracts again when it passes out of the slide and into the air and then again when it passes through the glass of the objective. This scattering of light does not affect the image quality at lower magnifications. However, at high magnification where you are viewing a very, very small section of the slide, the image becomes blurred and loses contrast. Immersion oil is used to minimize light scattering. Immersion oil has the same refractive index as glass. Light bends when it enters the glass and remains on the same path as it passes through the immersion oil and enters the objective. Since light scattering is decreased, the image is clearer and sharper and has better contrast.

### ✅ Tips for Success

- You will be using the same microscope all semester. Handle the microscope with care and clean it thoroughly at the end of each class.
- Do not force any of the microscope's adjustments and do not disassemble anything. If you have any problems with your microscope, inform your instructor at once.

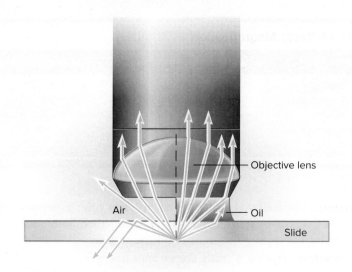

**Figure 1.3** Workings of an oil immersion lens.

- Make sure you place your slide on the stage with the coverslip up, closest to the objective.
- Always start focusing with the 4× objective.
- Always place the specimen in the center of your field of view before switching to the next higher power objective.
- If you are having trouble focusing on the slide, move the edge of the coverslip into your field of view and focus on that first.
- Bacteria are very small. Look for colored flecks of "dust" on your slide. Those flecks are globs of stained bacterial cells.
- Do not move the stage or focus adjustment knobs when applying oil!
- Use oil only with the 100× objective. **Do not use oil with the 40× objective.**

## Organisms

- *Bacillus cereus* or *Bacillus subtilis*
- *Staphylococcus aureus*
- Other slides as indicated by your professor
- *Spirillum volutans*
- *Saccharomyces cerevisiae* (yeast)

## Materials

- Microscope
- Immersion oil
- Lens paper
- Eyedropper or pipette
- Pond water (if available)
- Lens cleaner

## Procedure

### Viewing prepared slides

1. Obtain a microscope. Always carry the microscope upright, chest high and with both hands, one hand holding on to the arm and the other under the base. Place the microscope onto the laboratory bench.

If you need to reposition the microscope, lift the microscope carefully and reposition the microscope. **Do not drag your microscope across the laboratory bench.** Dragging the microscope can damage the lenses and illuminator. Keep your microscope away from the edge of the desk. Do not rest it on books or papers. Adjust your lab stool or chair for the proper viewing height.

2. Plug in the microscope and turn on the light. You should see light coming through the condenser and through the stage. If your microscope light does not turn on, make sure the cord/plug on the back is securely plugged in and the GFI (ground fault interrupter) trip on the power outlet is on.

3. Obtain a prepared slide from the slide tray in the common materials area.

4. Open the mechanical stage arm and slide the slide into place. Release the stage arm. The coverslip should be "up" or closest to the objective. If the slide is upside down, you will not see the bacteria under oil immersion.

5. Use the mechanical stage control knobs to position the specimen over the light passing through the stage.

6. Use the iris diaphragm slider to adjust the light. Start with your condenser diaphragm half open. As you increase magnification, you will need to increase the amount of light passing through your specimen. Viewing specimens at higher magnification requires more light.

7. Grab the nosepiece and rotate the 4× objective over the specimen. Always start focusing with the scanning objective. Do not move to the next higher objective if you do not have a sharp image at the current magnification. If you do not see the object at low power, you won't see it at high power either.

8. Grasp the coarse adjustment knobs and rotate the knobs to raise the stage to its highest point.

9. Adjust the oculars for your interpupillary distance.

10. Look into the oculars and use the coarse adjustment knob to slowly lower the stage. The slide will slowly come into focus. Look for colored, dustlike spots in your **field of view.** Use the fine adjustment knobs to sharpen the image.

11. Use the mechanical stage adjustment knobs to move the dustlike spots into the center of your field of view. If your image is not centered, it may not be in your field of view when you rotate to the next objective.

12. Rotate the 10× objective over the slide. Because your microscopes are **parfocal,** your specimen should still be visible and in relatively good focus. Use the fine adjustment knob to sharpen your image. You should not have to use the coarse adjustment knob. Center your specimen in the center of your field of view.

13. Rotate the 40× or 45× objective over the slide. Use the fine adjustment knob to sharpen your image. You should not have to use the coarse adjustment knob. Center your specimen in the center of your field of view.

14. To increase the available **working distance,** rotate the 4× objective over the slide. **DO NOT MOVE THE STAGE** or **THE MECHANICAL STAGE.** Add a drop of immersion oil directly to the slide.

15. Rotate the oil immersion lens into position over the slide. The objective should enter the oil. **Do not** move the stage down; the lens must be in the oil and very close to the slide. If you follow the correct focusing procedure, you will not break the slide.

16. Use the fine adjustment knob to sharpen your image. The **depth of field** is not very large at this magnification. As you perform fine focus adjustments, focusing up and down, individual cells will come in and out of focus quickly.

17. Record your results in your lab report.

18. Use this same procedure to observe all of the prepared slides and record your observations of these slides in your lab report.

19. All prepared slides must be cleaned (wipe with paper towel and lens cleaner) and returned to the correct slide tray.

## (Optional) Preparing and viewing a wet mount of pond water

1. Obtain a clean blank slide and coverslip.

2. Acquire a pipette or eyedropper.

3. Gently squeeze the bulb of the pipette and insert it into the container of pond water. Place the pipette near any solid material or noticeably chunky material in the sample. Release the bulb.

4. Squeeze the bulb gently to dispense **one** drop of pond water onto the clean slide. Too much liquid will make observing the specimen more difficult.

5. Place the plastic coverslip over the drop.

6. View the slide (see steps 4-13 of "Viewing prepared slides"). You should not use oil immersion with this specimen.

7. Draw in your lab report two different types of organisms from your pond water sample.

8. When you have completed your drawings, dispose of your slide and coverslip appropriately.

## Cleaning your microscope

1. The optical elements of your microscope should be cleaned only with lens paper and lens cleaner. Cloth or other types of paper will scratch the optical glass. Add a drop of lens cleaner to the lens paper and vigorously rub the oil immersion objective. Clean the ocular and other objective lenses with another piece of lens paper and lens cleaner.

2. Water and oil on the surface of the stage may be cleaned using a paper towel.

## How to Properly Store Your Microscope

1. Remove any slides from the stage. Clean and dry the stage with a paper towel.

2. Lower the stage and place the 4× objective over the stage.

3. Coil the power cord onto the cord supports on your microscope.

4. Return your microscope to its proper position in the storage cabinet.

## Results and Interpretation

Natural environments, like pond water or the microbiome in our body, contain billions of organisms from the various domains. This exercise should give you a glimpse of this diversity. You may want to review the exercises on microbial phototrophs and protozoa for some hints on the identity of your microbes. Visual and microscopic observation of suspect environmental conditions is just one of many techniques used to evaluate the health and safety of these environments.

Name _____

Date _____

# Use of the Microscope

## Your Results and Observations

In the circles below, draw the specimens you observed. Fill the circle with the specimen's image. You will need to exaggerate the organism's size. Include the magnification at which you observed the specimen. Label the images appropriately.

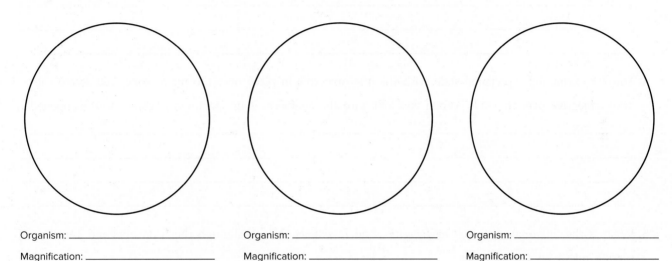

Organism: _____

Magnification: _____

Organism: _____

Magnification: _____

Organism: _____

Magnification: _____

## Interpretation and Questions

**1.** In this laboratory you used a compound light microscope. There are many other types of microscopes, including stereo/dissecting microscopes, fluorescent microscopes, phase contrast microscopes, and various types of electron microscopes. Use your textbook or other resources to determine which microscope should be used to visualize these specimens, and provide a brief explanation of your selection.

| Specimen | Microscope type | Why this microscope? |
|---|---|---|
| Whole mount of fly | | |
| Ebola virus | | |
| *Plasmodium* spp. (protozoan) | | |
| Muscle cells | | |

**2.** As the staff biologist for the public health department, you examine the pond water (case file) under the microscope and notice high levels of green algae. What do you tell the members of the community about the safety of their water? What other information do you need?

_____

_____

_____

_____

**3.** What is the purpose of using immersion oil with the 100× objective? (You may find it helpful to review the "Oil Immersion Microscopy" section and take a look at figure 1.3.)

_____

_____

_____

_____

**4.** You are examining a prepared slide with the 40× objective in place over the microscope. You rotate the 100× objective over the slide. When you look into the eyepiece, your specimen is gone! What happened?

_____

_____

_____

_____

**5.** What is the difference between _magnification_ and _resolution_? (A look through the "Concepts and Terms" section might be a good idea.)

_____

_____

_____

_____

# Tools of the Laboratory:
## Basic Techniques

# Aseptic Techniques

Three cases of bacterial meningitis in postpartum women were reported to the health department. All of these women had recently given birth in the same hospital, and all had received intrapartum spinal anesthesia. All were confirmed to have *Streptococcus salivarius* meningitis; one woman subsequently died. *S. salivarius* is a normal member of the human microbiota. Further investigation indicated a single anesthesiologist was present during the procedures of all three women, which involved spinal-epidural anesthesia. The anesthesiologist did not wear a mask during the procedure. Staff members reported that unmasked visitors to the operating room were common. *S. salivarius* was subsequently isolated from nasopharyngeal swabs taken from the anesthesiologist. The hospital in question instituted new infection-control procedures during spinal procedures, including the use of a mask and strict adherence to aseptic technique.

©Chris Ryan/AGE Fotostock RF

## LEARNING OUTCOMES

At the completion of this exercise, students should be able to

- demonstrate the ability to aseptically transfer bacteria between various media.
- describe and explain the steps involved in the aseptic transfer of an inoculum.
- describe the importance of aseptic technique in laboratory and clinical settings.

## Background

The ability to cultivate organisms of interest in the laboratory without contamination is critical for the study and characterization of those organisms. The method of handling microbes and materials in a way that minimizes contamination is called **aseptic** (*a* = "without," *sepsis* = "contamination") **technique.** Often, we transfer microorganisms from one medium to another for further study or to maintain cultures. This process of transferring a microbe from one medium to the next is called **inoculation,** and the sample being transferred is called an **inoculum.** The techniques described in this lab will be utilized throughout this semester; therefore, it is essential that you understand the basics of this technique to prevent contamination of the sample, others,

(a)

(b)

**Figure 2.1 Inoculating instruments.** The inoculating loop **(a)** is used for most of the transfer procedures done in this class. The inoculating needle **(b)** is used for techniques that require stabbing into the agar such as is done with the triple sugar iron (TSI) slant. ©Steven D. Obenauf

**Figure 2.2** Bunsen burner being used to sterilize inoculating instruments. ©Steven D. Obenauf

**Figure 2.3** Holding the tube, cap, and loop. ©Steven D. Obenauf

the environment, or yourself. These techniques require some manipulative skill and can be challenging when you are first learning how to handle the tools and the specimens. **Be persistent** and don't take any shortcuts; your efforts will result in reliable organism isolation and test results.

## Working with Inoculating Instruments

Two instruments are commonly used to transfer an inoculum: an **inoculating loop (figure 2.1a)** and an **inoculating needle (figure 2.1b).** The needle and the loop are made from Nichrome, a metal alloy that heats up very quickly.

These instruments are sterilized using a Bunsen burner flame **(figure 2.2)** or in a microincinerator. The flame should be blue, not yellow. The hottest part of the burner flame is just above the inner blue cone. To properly sterilize the loop, place the back of the loop wire, where it meets the handle, just above the inner blue cone. Hold the wire in place until it becomes red hot. Pull the rest of the wire slowly through the flame, making sure that it stays red hot the entire time. Alternatively, place the instrument in the flame, holding the instrument nearly vertically until the entire wire glows.

Once sterilized, do not place the loop or needle on the benchtop. Allow the loop or needle to cool slightly before putting it into a culture, or you may kill your bacteria. Do not wave the loop around to cool it. Do not blow on the loop. Do not touch the loop to check if it is still hot. Next, transfer your organisms (see "Removing an inoculum from a liquid culture"). When finished with your transfer, flame your loop again to incinerate any bacteria remaining on the tool. You should sterilize your loop immediately before and after every transfer. Your sterilized loop can then be stored in a test tube rack or storage area. Never lay it down on the counter or against the tubing of your Bunsen burner.

## Opening Test Tubes

Every time you open a test tube, the mouth of the tube must be sterilized. First, remove the cap using the **last two fingers** (pinkie and ring finger) of the hand holding the loop **(figure 2.3). Never** put the cap down on the benchtop. Pass the mouth of the tube through the flame two or three times. Don't hold the tube in the flame for a long time or the glass may crack. Tilt the tube to minimize the entry of any contaminants from the surrounding air. Before replacing the cap, make sure to flame the tube again. You will use a similar technique when opening bottles.

The following section is divided into six subsections. The first three subsections describe the procedures for removing **inoculum** from a broth, a slant, or a Petri dish. The next three sections describe how to inoculate a broth, slant, or Petri dish with that inoculum. In this activity, you will be transferring bacteria between different media using these techniques. Make sure you are reading the correct subsection as you work through the exercise.

### Removing an inoculum from a liquid culture

1. Select the stock tube from which you are going to remove your inoculum. Keep the tube upright and roll the tube back and forth between your palms several times to resuspend the organisms. **DO NOT** shake the tube up and down.

2. Place the tube back in the rack and pick up your loop. The loop is **ALWAYS** used when transferring organisms from a liquid medium.

3. Flame and hold your loop as just described.

4. Pick up the stock tube used in step 1.

5. Remove the cap using your pinkie and ring finger of the same hand that is holding the loop.

6. Tilt the test tube slightly and pass the mouth of the test tube through the Bunsen burner two or three times. Tilting the tube prevents the culture from pouring out and minimizes the surface area of the tube exposed to contaminants in the air.

7. Insert the loop into the broth. Remove a loop of culture **(figure 2.4).**

8. Flame the mouth of the tube again.

9. Recap the test tube and return the stock tube to the rack.

10. Transfer your inoculum to another medium.

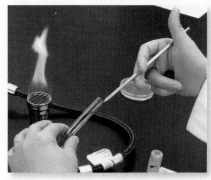

**Figure 2.4** Removal of inoculum from a broth culture. ©Steven D. Obenauf

## Removing an inoculum from a slant

1. Select a slant stock tube.

2. Remove the cap and flame the mouth of the tube as described previously.

3. Touch the loop (the loop should be in the hand holding the stock tube cap) to the surface of the slant. Keeping your loop hand still, pull the tube away from the loop **(figure 2.5a).** The loop should scrape a very small amount of culture from the slant. Your loop should be no more than one-quarter full.

4. Flame the opening of the tube again and replace the cap.

## Removing an inoculum from a plate

1. It is important to keep the Petri dish with your stock culture close to the Bunsen burner during the transfer. Take a few minutes to organize your materials for easy access.

2. Flame and hold your loop as just described.

3. Lift the Petri dish lid at about a 45-degree angle **(figure 2.5b). DO NOT** completely remove the lid. To minimize contamination from your surroundings, never leave a Petri dish open to the environment.

4. Gently rub the surface of the agar and remove an inoculum the size of a comma. You do not need to scrape the plate clean. Remember, the bacteria are growing on the surface of the agar. Try not to plow through the medium.

5. Remove the loop and close the Petri dish lid.

6. Transfer your inoculum to another medium.

## Inoculating a slant

1. Remove your inoculum from a stock culture. See the steps previously described.

2. Pick up an agar slant from the supply table. Label the slant with your group designation, the date, and the organism's name.

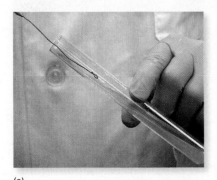
(a)

(b)

**Figure 2.5 Removing inoculum from solid media.** The removal of inoculum from a slant **(a)** and from a plate **(b)** is shown. The Petri dish lid is used to shield the plate from airborne contaminants. ©Steven D. Obenauf

3. Remove the cap using your pinkie and ring finger of the same hand that is holding the loop.

4. Tilt the test tube slightly and pass the mouth of the test tube through the Bunsen burner two or three times.

5. With a steady hand, insert the loop as far into the tube as you can without touching the sides or agar surface **(figure 2.6a)**.

6. Place the loop on the surface of the agar and then pull the slant away while keeping the loop in constant contact with the agar surface. This should produce a straight line inoculation on the slant.

7. Lift the loop before you get to the lip of the tube.

8. Flame the mouth of the tube and replace the cap.

9. Flame your loop.

## Inoculating a broth

1. Remove your inoculum from a stock culture. See the steps previously described.

2. Pick up a broth tube that you have already labeled with the organism's name or number.

3. Remove the cap using your pinkie and ring finger of the same hand that is holding the loop.

4. Tilt the test tube slightly and pass the mouth of the test tube through the Bunsen burner two or three times.

5. With a steady hand, insert the loop into the broth without touching the sides of the tube.

6. Rotate the loop handle to stir the bacteria into the medium. Don't worry if your comma-size inoculum does not come off the loop. Even with a slight jiggling loop-hand motion, you have introduced enough bacteria to see results.

7. As you remove the loop, touch it to the side of the tube to remove any excess liquid.

8. Flame the mouth of the tube and replace the cap.

9. Flame your loop.

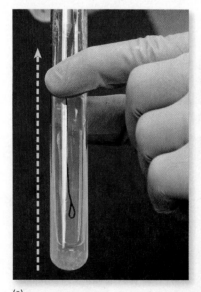 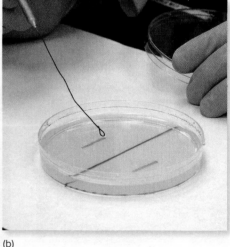

(a)                                      (b)

**Figure 2.6 Inoculating solid media.** Inoculating a slant from bottom to top **(a)** and inoculating a plate **(b)** (note that the lid should be held over the plate at a 45-degree angle—it has been moved so you can more easily see the plate).
©Steven D. Obenauf

## Inoculating a plate

Many of the experiments in this book are designed to test an organism's ability to grow on a specific medium or metabolize a particular compound. Spot inoculations will be used for these tests.

1. Acquire an uninoculated Petri dish from the supply table.

2. Use a marker on the bottom of the plate to divide it into halves or quadrants. Label the bottom of the Petri dish with your group designation, the date, and the organisms' names (do not write in the center of the Petri dish; write along the edge).

3. Remove your inoculum from a stock culture. See the steps previously described.

4. Lift the lid of the plate at a 45-degree angle to avoid any contamination **(figure 2.6b)**. Touch the loop to the surface of the agar and pull it in a straight line for about a centimeter or make an "S" pattern. Do not puncture the agar. Remove the loop and replace the lid. Flame your loop. Isolation streak inoculations will be described in later exercises.

### ✔ Tips for Success

- Make sure your benchtop is organized and not cluttered. Have all the required materials within easy reach.
- Keep the Bunsen burner in the center of your bench, not near the edge.
- Know where the burner is at all times and be careful when reaching across the bench. Do not leave your Bunsen burner unattended.
- When transferring bacteria, sterilize your loop before and after transfer.
- After sterilizing your loop, allow the loop to cool before transferring bacteria.
- Never place the cap of a tube on the benchtop. Test tube caps are a source of contamination. Protect yourself and your workspace by always holding the caps in your hand.
- Concentrate on what you are doing. Avoid conversations with your lab partners. Always be aware of where you are in the process. For example, did you flame your loop and cool it before removing your inoculum? Did you flame the tube before replacing the cap?
- Shoulder-length or longer hair needs to be constrained.

### Organisms

- *Escherichia coli* (broth)
- *Staphylococcus aureus* (broth)
- *Bacillus cereus* (broth)
- *Micrococcus luteus* (broth)
- *Serratia marcescens* (plate)

### Materials (per Team of Two)

- 2 nutrient agar slants
- 1 tryptone soy broth
- 1 nutrient agar plate
- Inoculating loop
- Bunsen burner and striker or microincinerator
- Test tube rack, storage can

### Procedure

#### Period 1

1. Each group will inoculate two slants, one broth, and one nutrient agar plate. Everyone needs to practice these techniques, since they will be used throughout the rest of the course. Divide the work, so that each partner inoculates media. Before starting, label all your tubes. Remember to write on the glass test tube, not the cap.

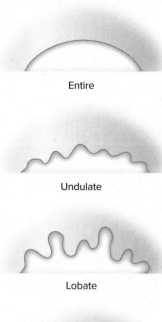

Entire

Undulate

Lobate

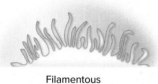

Filamentous

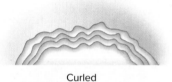

Curled

**Figure 2.7** Colony margin or edge characteristics.

2. Select two of the stock organisms in broth culture; inoculate one nutrient agar slant for each organism (see "Removing an inoculum from a liquid culture" and "Inoculating a slant" for detailed instructions). Each student will transfer one organism from a stock broth tube to a nutrient agar slant. Make sure to label your tubes.

3. Aseptically transfer a small amount of bacteria from the *Serratia marcescens* plate (see "Removing an inoculum from a plate" for detailed instructions) and inoculate a tube of tryptone soy broth (see "Inoculating a broth" for detailed instructions). Do not place the Petri dish lid on the benchtop; remember to keep it in your hand at all times.

4. Using two of the four organisms in broth culture, inoculate a nutrient agar plate. Each team member will inoculate one-half of the Petri dish (see "Inoculating a plate" for detailed instructions) with one organism.

5. Place your inoculated tubes in a can or storage container. Label the container with your group name, the class time, and the date. Place the container in the incubator. Place your Petri dish into the incubator, lid side down.

6. Return or discard your stock cultures as instructed.

## Period 2

1. Observe your slants, plate, and broths for growth. Do you have good growth? Do you see any contamination?

2. Look at the edge or margin of the growth on your slants and compare it to the diagrams in **figure 2.7.** Is it entire, undulate, lobate, curled, or filamentous? Record your results.

3. Carefully hold up your broth tubes **(don't tip them!)** and examine the growth in your broth. Are the tubes uniformly cloudy **(turbidity),** or are there flecks or chunks of bacterial growth scattered throughout the tube **(flocculent)?** Has growth settled onto the bottom **(sediment)?** Tilt your broth tubes to about a 45-degree angle and look at the surface of your broth. Is there a film of bacteria covering the surface (a **pellicle**) or growth only around the edge (a **ring**) or no surface growth? Record your results in your laboratory report.

## Results and Interpretation

As can be seen from the case file given at the beginning of this activity, the correct application of aseptic technique is important not only in the laboratory but also in the hospital or clinical environment. Bacteria are everywhere and on everything. Proper usage of aseptic technique will not only prevent contamination of your experimental media but could also have prevented the death of the patient in this case file.

When examining the results of your inoculations, look for consistency in the growth of your organisms. Are all of the colonies the same color, and do they have the same transparency? Do all of the colonies on a plate or slant have the same elevation and edge characteristics? If so, you probably aseptically transferred your inoculum. Congratulations! If you notice different-colored colonies or colonies with different growth characteristics, then somewhere during the procedure a contaminant was introduced. Think carefully about where in the procedure the contaminant could have been introduced. You may want to refer back to these procedures the next time you need to transfer bacteria. Practice makes perfect!

Name _____

Date _____

# Aseptic Techniques

## Your Results and Observations

**1.** Describe and draw the growth of your organisms on the nutrient agar slants; be sure to include the organisms' names.

| Organism | Describe (and draw if needed) the form or margin of the growth (see which part of figure 2.7 is most similar to your growth). |
|----------|---|
| *Staphylococcus aureus* | |
| *Bacillus cereus* | |
| *Escherichia coli* | |
| *Micrococcus luteus* | |

**2.** Describe the form of growth observed in the tryptone broth tubes of your organisms (terms you might use include *turbidity, flocculent, sediment, ring,* and *pellicle*).

_____

_____

_____

_____

**3.** Describe and draw the growth of your organisms on the Petri dish; be sure to include the organisms' names.

_____

_____

_____

_____

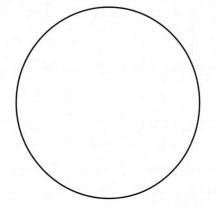

**4.** Would cultural characteristics like colony morphology be of value to a clinical microbiologist? How could they be used?

_____

_____

_____

_____

**5.** Were your cultures consistent in appearance? How could you confirm your culture was a pure culture? Where during the process of aseptic transfers could contaminants be introduced?

_____

_____

_____

_____

**6.** Jane was very careful in performing her Petri dish inoculation. When she examined her plate during the next lab period, a fuzzy white colony was growing on her plate, but not along her inoculation streak. What happened? Where did this colony come from?

_____

_____

_____

_____

**7.** Jeffrey, Jane's partner, inoculated a tryptone broth tube with _E. coli_. When he examined his broth tube during the next lab period, there was no growth at all. What could have gone wrong with his experiment?

_____

_____

_____

_____

**8.** In the case file, the organism responsible for the patient's death is a normal member of the human microbiota. How do you explain it causing the death of this patient?

_____

_____

_____

_____

**9.** What is the connection of the events described in the case file and today's laboratory activity on aseptic technique?

_____

_____

_____

_____

EXERCISE

# 3

# Isolation Streak Plate

## CASE FILE

Three patients who had been treated at the same pain clinic near the coast of New Jersey were diagnosed with severe infections by methicillin-resistant *Staphylococcus aureus* (MRSA), including meningitis, epidural abscesses, and sepsis. All had received spinal steroid or anesthetic injections for back pain. Investigation found that the contrast medium they received to guide those injections was taken from a single-dose vial (intended for use for only one patient) and given to multiple patients. It was determined that personnel did not wear face masks while performing spinal injections. Nasal swabs were collected from clinic medical providers and staff members who were involved with the preparation or administration of injections. Further procedures were done to determine if they or the contrast media were the source of the MRSA infections. In order to do DNA testing, serotyping, or other tests, the different bacteria present in clinical or environmental samples must be separated from each other.

Source: CDC/Janice Haney Carr

## LEARNING OUTCOMES

At the completion of this exercise, students should be able to

- explain the purpose of the isolation streak plate method.
- demonstrate how to perform an isolation streak plate procedure.
- examine a plate and determine if isolated colonies were obtained and whether it was a success.

## Background

Samples from patients or the environment are usually mixtures of a number of different bacteria. Most laboratory work done with bacteria requires pure cultures of a single organism, so the members of mixed populations must be separated from each other in some way. This can involve the use of spread plates, pour plates, or isolation streak plates.

The **isolation streak plate** is a widely used method to separate individual bacteria from mixtures. There are a number of somewhat different isolation streaking patterns, but they all have the same purpose. The goal of the streaking pattern is to isolate individual bacteria from the mixture as a result of having fewer and fewer bacteria as you continue around the plate. It is almost as if you are painting one long, continuous stripe of

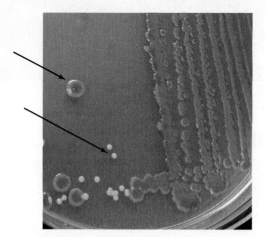

**Figure 3.1** Isolated colonies. ©Steven D. Obenauf

bacteria on the surface of the plate. Eventually, the concentration of bacteria gets so low that a single bacterium or a small group of cells comes off the loop to form single **isolated colonies** on the surface of the plate **(figure 3.1).** Usually, each **colony** arises from a single bacterium that divides repeatedly over time to form a mass of millions of cells. Therefore, all the bacteria in the colony should be the same. These single colonies can then be picked off the plate with your inoculating loop for further testing.

The technique we will use is sometimes known as the quadrant technique. This is because the first streak will cover half of the plate, while the next two will each cover half of the rest (second half) of the plate. It is important to know how to do this well. In a later exercise, the streak plate will be the first step in working on your unknowns.

## ✔ Tips for Success

- Try to follow the pattern as closely as possible. Creativity is a good thing in some situations, but not when doing an isolation streak plate!
- You may find it helpful to practice the pattern before making your plate by drawing it with a Sharpie on the lid of your Petri plate or with pencil on a sheet of paper.
- Touch your loop gently to the agar. Try not to dig in or cut the surface of the plate.
- When streaking your plate, hold the lid at an angle covering the plate or you may add some extra microbes to your mixture.
- Keep your plate in the incubator lid side down. If the lid is on the top and water from condensation drops onto the surface of your plate, there's a pretty good chance your streak will be ruined.

## Organisms (in Broth Cultures)

- Mix A (premade: contains *Escherichia coli* and *Serratia*)
- Mix B (premade: contains *Micrococcus* and *S. aureus*)

## Materials (per Student)

1 nutrient agar plate
Optional: blood agar can be utilized in place of nutrient agar

Inoculating loop

## Procedure

### Period 1

**1.** Each student should do one plate.

**2.** Use either mixture (but only one).

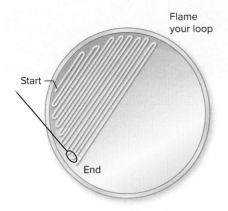

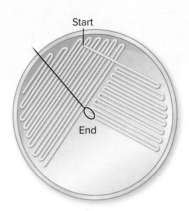

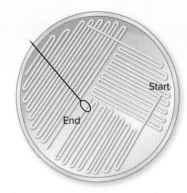

Flame your loop and dip it into the unknown mixture. Try to make your streaks parallel and close together. Your first streak area should cover about half of the plate.

Flame your loop. *Do not* dip into the unknown mixture again. Drag your loop across one side of your first streak area. Make your streaks close together and try not to cross back into the first streak area. Your second streak area should cover a quarter of the plate.

Do not flame your loop before streaking the third area. Drag your loop across one side of your second streak area. Make your streaks close together and try not to cross back into the first or second areas. Your third streak area should cover a quarter of the plate.

**Figure 3.2** Steps in streaking a plate.

3. Using your loop, follow the streaking pattern indicated by **figure 3.2** (alternative patterns may be provided by your lab professor or lecture textbook).

4. Put your plate in the incubator, lid side down.

## Period 2

1. Observe and draw your plate.

2. Look for isolated colonies. Determine if you separated the two bacteria in your mixture similar to the results shown in **figure 3.3a** or were not successful **(figure 3.3b** and **c)** and may need to try again.

## Results and Interpretation

Figure 3.3a shows a successful plate. The **first streak area** shows heavy, confluent growth. In the **second streak area,** you can see that the growth is starting to thin out and some individual colonies are starting to appear. Finally, in the **third streak area,** thinner growth and isolated colonies appear. Figure 3.3b shows a plate where there was no flaming or poor flaming after area 1, and figure 3.3c shows a plate where an improper pattern was used.

(b) No or poor flaming after area 1

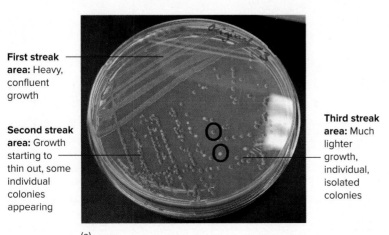

**First streak area:** Heavy, confluent growth

**Second streak area:** Growth starting to thin out, some individual colonies appearing

**Third streak area:** Much lighter growth, individual, isolated colonies

(a)

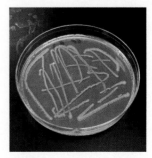

(c) Improper pattern

**Figure 3.3** Successful and unsuccessful isolation streak plates. ©Steven D. Obenauf

# NOTES

Name _____

Date _____

# Isolation Streak Plate

## Your Results and Observations

**Draw** your plate. You may also photograph it with a digital camera or mobile device.

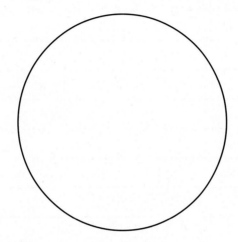

## Interpretation and Questions

1. Was your plate successful? Why or why not (you may want to review the Background section and look at figures 3.1 and 3.3)?

_____

_____

_____

_____

2. When streaking a plate, the loop is pulled from streak area 1 to streak area 2 and then from streak area 2 to streak area 3. Why should the loop not be pulled through streak area 1 through streak area 3?

_____

_____

_____

_____

# Microbial Phototrophs:
# Algae and Cyanobacteria

## CASE FILE

The Florida Department of Health was notified by a local hospital that eight patients had been seen with one or more of the following symptoms: cramps, nausea, dizziness, vomiting, diarrhea, and chills and sweats. The onset of symptoms occurred within hours of eating at the same local restaurant and persisted for 12 to 24 hours. During the next 5 days, these same individuals reported pruritus of the hands and feet, unusual sensations, including tingling of the skin; and muscle weakness. Stool and vomitus samples were tested for the presence of *Salmonella, Vibrio, Shigella, Yersinia,* and *Campylobacter,* but results were negative. Sanitation standards at the restaurant were acceptable and not thought to be the cause of the outbreak. Health officials suspected that these illnesses were ciguatera poisoning, resulting from the consumption of contaminated amberjack. An investigation revealed that the amberjack was shipped from Key West and distributed to restaurants and grocery stores in northern Florida and Alabama. Nine more people from the distribution areas reported similar illnesses after eating amberjack.

©David M. Phillips/ Science Source

Ciguatera toxin is a naturally occurring toxin produced by the dinoflagellate *Gambierdiscus toxicus.* This organism is common to the tropical waters of the Pacific Ocean and the Caribbean. Most outbreaks occur in Florida and Hawaii. The lipid-soluble toxin bioconcentrates up the food chain. Toxin-producing organisms are eaten by herbivorous fish, which in turn are eaten by larger carnivorous fish. The larger, predatory reef fish contain the highest levels of ciguatera toxin. This toxin is not affected by cooking or freezing. It is tasteless, colorless, and odorless.

### LEARNING OUTCOMES

At the completion of this exercise, students should be able to
- list the major groups of microbial phototrophs.
- recall the environmental and clinical significance of these organisms.
- describe the major characteristics, structures, and by-products of microbial phototrophs.

## Background

The microbial world is made up of organisms from different domains (kingdoms). They have in common a microscopic form at some point in their life cycle. Today's lab will examine some of the microbial phototrophs. These organisms belong to different kingdoms yet have in common the ability to perform oxygen-evolving photosynthesis. You will view and draw members of the prokaryotic cyanobacteria, formerly called the blue-green algae and some of the eukaryotic algae.

### Cyanobacteria

The classification of the cyanobacteria has always been a challenge to scientists. They have in the past been classified as plants, protozoa, and bacteria. Because they are prokaryotic (their DNA is contained within a region of the cell called the nucleoid), the cyanobacteria are now classified as bacteria and belong to the domain Bacteria. Like other prokaryotic members of this domain, they lack a nucleus and other double-membrane-bound organelles. They do possess an internal membrane system associated with their photosynthetic apparatus. Cyanobacteria derive their cellular energy from oxygen-evolving photosynthesis and utilize chlorophyll *a*, the same pigment used by higher plants, to capture light energy. In addition to chlorophyll *a*, they contain accessory pigments called phycobilins. These pigments are frequently used when developing diagnostic tests for clinical use.

Cyanobacteria grow in soils, freshwater, and saltwater and are more numerous and varied than any other photosynthetic bacterial group. They exist as single cells; filaments (strands) of cells that look like chains of beads; or flat, ribbonlike colonies of cells. They can be an important environmental concern. They can have either a detrimental or positive effect in an environment. Cyanobacterial populations respond quickly to changes in nutrient levels in aquatic systems and are frequently involved in algal blooms in polluted environments. Their rapid growth under these conditions leads to a tremendous increase in algal numbers, which in turn depletes available nutrients. As nutrient levels decrease, the algae start to die. The bacterial breakdown of the dead algae can cause dissolved oxygen levels in the water to plummet, negatively impacting lakes, rivers, and canals. Some blooms are associated with the production of toxins, which can cause respiratory or gastrointestinal problems in people. They also are beneficial to the health of soils by restoring nitrogen levels through the process of nitrogen fixation. Cyanobacteria are well known for their involvement in symbiotic relationships. Some of the earliest forms of life noted in the fossil record (stromatolites) are thought to be a symbiotic relationship between cyanobacteria and fungi. Lichens are important environmental indicators and are the result of an association between cyanobacteria and fungi.

Morphology is still the mainstay of identification of the cyanobacteria. They are larger and more complex than many other bacteria. The following is a list of structures and terms associated with the cyanobacteria. Keep them in mind and try to identify these as you examine your slides.

**Akinete (figure 4.1*b*).** Akinetes are reproductive structures. They are large, usually about two to three times larger than the average cell in a filamentous strand. The akinete will have a thicker wall and a grainy, mottled appearance.

**Heterocyst (figure 4.1*a*).** Heterocysts are thick-walled cells that are responsible for nitrogen fixation. Nitrogen is an essential element. All living things require nitrogen to synthesize nucleic acids and proteins. Nitrogen, however, is not always freely available in a form usable by living organisms. Nitrogen gas is plentiful in our atmosphere

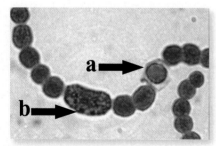

**Figure 4.1** *Anabaena. Anabaena* is a genus of filamentous cyanobacteria. A thick-walled heterocyst is shown at the arrow tip **(a).** An akinete is also shown at the arrow tip **(b).** (Magnification: 1,000×)
©Susan F. Finazzo

(about 78% of the atmosphere is $N_2$); however, only a few organisms can utilize gaseous nitrogen. Nitrogen-fixing cyanobacteria absorb nitrogen gas and produce the enzyme nitrogenase. Nitrogenase converts gaseous nitrogen into an organic form. This enzyme is inactivated by oxygen, so heterocysts are thick-walled to inhibit the diffusion of oxygen, and these specialized cells lack chlorophyll. Heterocysts are thick-walled, very round, and have a smooth, even coloration.

**Trichome.** Trichomes are strands of filamentous cyanobacteria contained within a sheath. Short trichomes that break away from a mass and form a new "colony" are called hormogonia.

**Sheath.** Some of the cyanobacteria produce a covering called a glycocalyx, or sheath. The sheath varies in composition and thickness. The sheath is an essential feature for gliding motility. How gliding motility occurs is still a mystery. However, cells that have a sheath have the ability to glide when in contact with a solid surface.

### ✓ Tips for Success (Cyanobacteria)

- These slides can be a little thick and the specimen may be in more than one plane of focus. You may need to use the fine focus knob to see an entire strand.
- Try to identify all of the unique structures just mentioned.
- Share the slides with your partner. After you focus and record your observations of a slide, trade places with your partner.

### Organisms (Cyanobacteria)

Prepared slides:

- *Anabaena*
- *Oscillatoria*
- *Rivularia*

### Procedure (Cyanobacteria)

1. View all slides of cyanobacteria using oil immersion or 400×. Record your observations in your lab report.
2. Clean your slides and return them to the correct tray after viewing.

## Algae

The green algae are unicellular (such as *Chlamydomonas*) or multicellular (such as *Spirogyra*) eukaryotic members of the kingdom Plantae. Some algae have been classified as protists (*Peridinium* is an example you will look at and a relative of the causative organism in the case file). Algae are a diverse group and can be found in freshwater, saltwater, and soils. They are classified by their photosynthetic pigments, form of reproduction, and motility and the composition of their cell walls.

Algae are **eukaryotic (figure 4.2).** They possess a distinct nucleus and a chloroplast containing their photosynthetic apparatus. All of the algae contain chlorophyll *a*. In addition, they may contain another form of chlorophyll (*b, c,* or *d*). They may be nonmotile or move by flagella or by gliding motility.

Algae are environmentally and economically important. As the primary producers in aquatic ecosystems, they are the basis for food chains. Oxygen, a by-product of photosynthesis, is essential for many forms of life. The vast majority of the oxygen in our atmosphere is produced by marine algae. They are an important food source for both animals and people. Agar, the solidifying agent used in media, is an algal product. Algae also produce

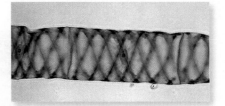

**Figure 4.2** *Spirogyra. Spirogyra* is a filamentous green alga. The spirals crisscrossing the cell are the chloroplasts. The nucleus is evident in the center of the cell.
(Magnification: 400×) ©Susan F. Finazzo

33

(a)

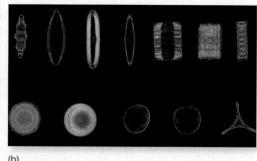
(b)

**Figure 4.3** Diatoms. Image **(a)** is a light micrograph of *Synedra*, a genus of diatom. Image **(b)** shows just a few of the many frustule shapes observed in diatoms. (Magnification: 400×) ©Susan F. Finazzo

carrageenan–a common additive in soups, ice cream, and other food and pharmaceutical products.

The **diatoms** are a group within the algae. They are a common component of phytoplankton. Diatoms exist as single cells or in colonial forms. They have very distinctive and sometimes elegant shells known as frustules **(figure 4.3).** The frustule is composed of silicon dioxide, the same element that makes up sand and glass. The frustule is made in two halves that fit together like a shoe box and lid. They are used as indicator organisms of environmental conditions both in present day and past environments.

The **dinoflagellates** are unicellular algae that are commonly found in phytoplankton. They are primary producers and an important food source in warm, marine tropical waters. There are many species of dinoflagellates–most are marine and most use flagella to move. The dinoflagellates have thecae, or stiff, armorlike wall structures. The thecae are composed of cellulose and are found within the plasma membrane. *Peridinium* is a planktonic dinoflagellate **(figure 4.4).**

The shape and orientation of the thecae produce elaborate and unique shapes. Dinoflagellates can exhibit phosphorescence–that is, they glow in the dark when disturbed or stimulated. They are also responsible for some algal blooms. Some of these blooms release toxins, like red tides, into the water, which can have lethal effects on marine animals and impact human health. Ciguatera fish poisoning results from bioaccumulation in certain fish of a toxin produced by another dinoflagellate, *Gambierdiscus toxicus*.

While it is not usually fatal, the symptoms can be quite severe. Dinoflagellate neurotoxins are also associated with an illness known as paralytic shellfish poisoning. It is caused by eating contaminated mussels, oysters, or clams that were harvested when there were high levels of toxic dinoflagellates in the water.

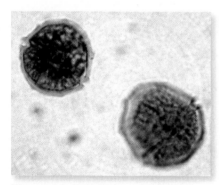

**Figure 4.4** *Peridinium. Peridinium* is a motile (flagella) dinoflagellate found in both fresh and salt waters. Notice the armorlike cellulose plates that compose the thecae. (Magnification: 1,000×) ©Susan F. Finazzo

|  | Bacteria Cyanobacteria | Algae Diatoms | Green algae | Dinoflagellates |
|---|---|---|---|---|
| Nucleus | No | Yes | Yes | Yes |
| Color | Yellow to brown | Yellow-brown | Green | Green |
| Movement | Gliding, flagella | Gliding | Nonmotile | Flagella |
| Cell wall composition | Glycocalyx–varied | Silica, with markings | Cellulose, rigid | Starch/cellulose |
| Produce $O_2$ | Yes | Yes | Yes | Yes |

## ✅ Tips for Success (Algae)

- *Chara* is a very large specimen. Use low power only.
- The *Synedra* slide can be difficult to see. Use low light. *Synedra* looks like shards of glass. The mixed diatom slide will have many broken fragments in it. Try to find the intact shells.
- *Chlamydomonas* is very small. Use low power to scan the slide for faint spots of color. Cells tend to stain in clusters.
- Share the slides with your partner. After you focus and record your observations of a slide, trade places with your partner.

## Organisms (Algae)

Prepared slides:

- *Spirogyra:* filamentous green alga; note spiral chloroplasts; 400×
- *Chlamydomonas:* single-celled green alga; 1,000×
- *Synedra* or diatom survey slide or diatomaceous earth slide; 40× or 100×
- *Peridinium:* dinoflagellate; 1,000×
- (Optional) *Chara:* 40×

## Procedure (Algae)

View all slides. Use the preceding materials section to determine the appropriate magnification to use. Some of the slides are very thick mounts. If you attempt to view them using the oil immersion lens, the coverslip may crack.

1. Record your observations in your lab report.
2. Clean your slides and return to the correct tray after viewing.

# Results and Interpretation

As you look at the prepared slides of cyanobacteria, look for the similarities between them. As you view at the various algae, look for what makes each of them unique. Although the slides you are looking at may contain only one or a few organisms, in the environment in which photosynthetic microbes live, they are part of a diverse mixture made up of hundreds or thousands of different members.

# NOTES

# Microbial Phototrophs: Algae and Cyanobacteria

## Your Results and Observations

In the circles below, draw your specimens; indicate the magnification. Label any distinctive structures. Identify the organism as an alga, a diatom, a dinoflagellate, or a cyanobacterium.

Organism: _____

Magnification: _____

Organism: _____

Magnification: _____

Organism: _____

Magnification: _____

Organism: _____

Magnification: _____

Organism: _____

Magnification: _____

Organism: _____

Magnification: _____

Organism: _____

Magnification: _____

Organism: _____

Magnification: _____

**1.** List at least three examples of how microbial phototrophs are important clinically and environmentally.

_____

_____

_____

_____

**2.** The toxin described in the case file caused dizziness, nausea, and muscle cramping. What does this suggest about its mode of action (i.e., what system does this toxin directly affect)?

_____

_____

_____

_____

# Protozoa

©micro_photo/Getty Images RF

Water quality can be tested using physical methods, chemical methods, and biological methods. **Bioindicators** are organisms that can show the effects of pollutants on an ecosystem. These include certain plants, animals (such as microinvertebrates), and protozoa. Observing protozoa such as *Paramecium* can be used to examine the quality of the water they live in, the potential for that water to cause disease, and the health of an ecosystem.

## LEARNING OUTCOMES

At the completion of this exercise, students should be able to

- describe the four major groupings of protozoa based on their means of movement.
- locate and identify three examples of free-living protozoa: *Amoeba*, *Euglena*, and *Paramecium*.

## Background

Protozoa are a diverse group of single-celled eukaryotic microorganisms that live in freshwater or saltwater. You've probably been swimming with them a few times! They use several different methods for moving, including flagella, cilia, undulating membranes, and amoeboid motion. Their internal structure is quite complex. Many, including the ones we will look at today, are predators of smaller organisms. Quite a few protozoa, in turn, serve as a food source for larger organisms such as invertebrate animals. Some, such as *Euglena* and dinoflagellates, are photosynthetic (we look at some examples

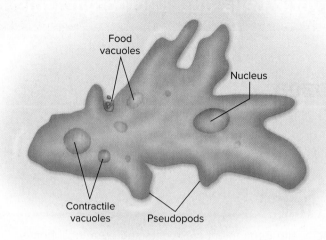

**Figure 5.1** Structure of *Amoeba.*

in the algae lab). The vast majority of protozoa play beneficial roles in ecosystems. A very small fraction of protozoa cause disease. We will look at the causes of malaria, giardiasis, sleeping sickness, and some other diseases caused by protozoa in the parasitic protozoa lab (exercise 29).

The classification of protozoa has changed a number of times over the years. For many years, they have been grouped together in the kingdom Protista. Recent evidence suggests that they are a group of phyla that do not all belong in one kingdom. Both the number of phyla and their names have been revised as new information has come to light about their relationships. A main factor (which has *not* changed) in how we classify protozoa is their method of movement. The four major groupings are those that use **flagella,** those that use **cilia,** those that use **amoeboid motion,** and parasites with **nonmotile** adult forms (sometimes known as sporozoa). In this lab, we look at three representative protozoa: *Amoeba proteus,* which moves by amoeboid motion **(figures 5.1 and 5.3),** *Paramecium caudatum,* which moves by means of cilia **(figures 5.2 and 5.4),** and *Euglena,* which moves by means of flagella **(figure 5.5).**

## ✅ Tips for Success

- For both of the prepared slides, start with your 4× objective to locate where the organisms are on the slide before going to a higher magnification.
- If live protozoa are available, sample by taking a dropperful of organisms from the bottom of the container or from the medium near the food source. Some of the organisms settle to the bottom, so you will see a better variety of them there.

## Organisms

*Amoeba proteus*

*Paramecium*

*Euglena*

Live protozoal mix (if available)

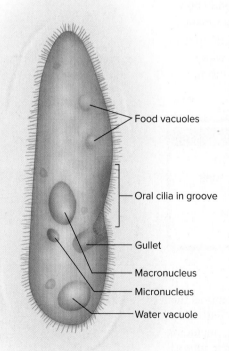

**Figure 5.2** Structure of *Paramecium.*

## Materials

- Dissecting microscopes (optional)
- Depression slides
- Plastic droppers
- Coverslips
- Methylcellulose solution

## Procedure (Prepared Slides)

1. Observe *Amoeba* by locating them on the slide first at 40× and then going to 100× magnification **(figure 5.3)**. Look for the major structural features, including the pseudopodia and the nucleus. The cells may be stained a variety of colors (living amoebas are actually colorless).

2. Observe *Paramecium* by locating them on the slide first at 40× **(figure 5.4a)** and then going to 100× or 400× magnification **(figure 5.4b)**.

3. Observe *Euglena* by locating them on the slide first at 40× and then going to 100× or 400× magnification.

4. Draw your *Amoeba, Euglena,* and *Paramecium* slides or photograph them if you are able.

## Optional Procedure (Live Mix)

1. To view the live mixed culture, make a wet mount by placing a drop of the culture on a depression slide. Use a standard or dissecting microscope **(figure 5.6)** to observe the culture. This may have been done for you as a demonstration. Dissecting microscopes have lower magnification than a standard light microscope, which can be helpful when observing moving organisms.

2. To look at them with your own microscope, place a drop from the bottom of the container into the well of a depression slide. Put a coverslip over the well.

3. If you use your own microscope, use 40× or 100× magnification to look at the live organisms. If the protozoa are moving too quickly to see, make a ring of methylcellulose around the well of the depression slide with a toothpick. Mix the methylcellulose into the drop of culture. If the organisms are not moving too quickly, do not use methylcellulose.

**Figure 5.3** *Amoeba* with multiple pseudopodia. (Magnification: 100×)
©Steven D. Obenauf

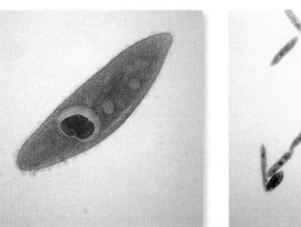

**Figure 5.4** *Paramecium.* (Magnification: 400×) Visible macronucleus, cilia, and food vacuoles. ©Steven D. Obenauf

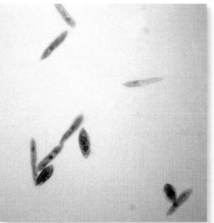

**Figure 5.5** *Euglena.* (Magnification: 100×)
©Steven D. Obenauf

**Figure 5.6** Dissecting microscope. ©Steven D. Obenauf

**4.** When you are done, please **wash and return the depression slide.**

**5.** Discard any coverslips or toothpicks in the sharps container.

## Results and Interpretation

As you look at the prepared slides of *Amoeba, Euglena,* and *Paramecium,* some perspective will be helpful. Although the slide you are looking at may contain only one organism, in the environment where these organisms live, they are part of a diverse mixture made up of hundreds or thousands of different members. If you look at a live culture, it is quite possible that it will have more than one organism in it.

Name _____

Date _____

# Protozoa

## Your Results and Observations

Draw or photograph with your mobile device the slides you observed.

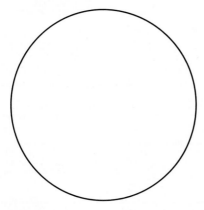

*Amoeba* Magnification: _____

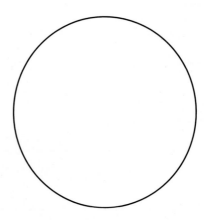

*Paramecium* Magnification: _____

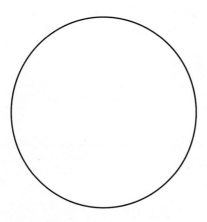

*Euglena* Magnification: _____

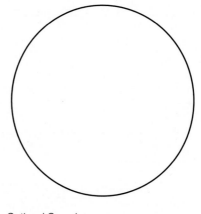

Optional Organism: _____

Magnification: _____

1. If a community is concerned about finding out the impact on the quality of water in a local river of nearby hydrologic fracturing (fracking) for gas, why might *Paramecium* or *Euglena* play a role in this process?

_____

_____

_____

_____

# Eukaryotic Cells and Microorganisms

# Fungi

## CASE FILE

A patient suffering from aortic insufficiency received an aortic valve allograft (transplant from an unrelated person). No immediate postoperative complications were noted and the patient was discharged 5 days postsurgery. Ten days later, he was readmitted to the hospital with a fever of 104°F (40°C), diarrhea, abdominal tenderness, and nausea. His white blood cell count was 7,000 per cubic milliliter (7,000/ml³). Blood cultures were positive for *Candida albicans*. A chemotherapeutic course of amphotericin B and 5-fluorocytosine was implemented. His physician suspected the patient had fungal endocarditis from a contaminated graft. A transesophageal echocardiogram confirmed separation of the aortic valve from the septum, and intramyocardial abscesses and vegetations around the suture line. The allograft was replaced. Seven days after the valve was replaced, the patient was fever-free. The surface of the suspect valve was cultured and the presence of *Candida albicans* confirmed.

Although fungal endocarditis associated with valve replacement is relatively rare, it can be a fatal complication. Unfortunately, *Candida* is a common cause of other types of hospital-acquired infections. It is the fourth most common cause of systemic hospital-acquired infections.

Courtesy of Danny L. Wiedbrauk

## LEARNING OUTCOMES

At the completion of this exercise, students should be able to
- list the major groups or phyla of fungi.
- differentiate between yeasts and molds, macroscopically and microscopically.
- recognize the vegetative and reproductive structures of molds.
- distinguish between buds, hyphae, and pseudohyphae.

## Background

From a completely human standpoint, members of the kingdom Fungi are both friend and foe. There are fungal pathogens. *Candida albicans* **(figure 6.1),** a yeast, is the causative agent of candidiasis, thrush, and other diseases. Fungi also cause the infections commonly called athlete's foot

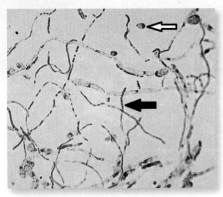

**Figure 6.1 *Candida albicans.***
*Candida* is a yeast in the phylum Ascomycota. In this image are distinct individual yeast cells (white-filled arrow); *Candida* also produces pseudohyphae (black arrow). (Magnification: 1,000×)
©Susan F. Finazzo

and ringworm. Beneficial fungi and their metabolic processes provide us with many food products, including cheese, bread, mushrooms, and alcoholic beverages. They are also responsible for the production of the vast majority of antibiotics currently used in medicine today. Agriculturally, fungi are very important in the biological control of nematodes and as plant partners (symbiosis) in lichen and mycorrhizal relationships. These last contributions are discussed more thoroughly in the soil microbiology exercise. Finally, they are critically important as decomposers. Fungi have the ability to degrade and recycle many complex organic molecules that would otherwise never break down.

The study of fungi is called *mycology*. Medical mycology is the study of fungal infections in humans. Fungi produce superficial infections such as athlete's foot. They also cause systemic infections such as cryptococcal meningitis, which infects nearly 1 million people each year. The incidence of fungal infections has been increasing annually worldwide. These increases result from new environmental fungal agents being associated with infections and with an increase in antibiotic/antifungal resistance in known pathogenic agents. Compounding these concerns is the difficulty in recognizing fungal infections. Many of these infections remain undiagnosed. Even when diagnosed, fungal infections are challenging to treat.

The members of this kingdom are eukaryotic. They grow either as unicellular or multicellular forms. Some genera can actually grow either as single cells or in the multicellular form, depending on the environmental and nutrient conditions. These organisms exhibit dimorphism–that is, they can exist in two forms, either molds or yeasts. The unicellular fungi are called **yeasts (figure 6.2).** The yeasts grow as individual cells, and their colonial form is very similar to a mucoid bacterial colony.

The multicellular forms are called **molds.** Mold growth is filamentous. The filaments are long, thin strands, which individually are called **hypha** (plural, *hyphae*). The cell walls and filament walls are composed of chitin. The hyphae are classified either as septate (have walls separating cells) or aseptate (without cell walls). In fungi with aseptate hyphae, the cytoplasm and nuclei can flow freely along the length of the filament. In fungi with septate hyphae, the nuclei are restrained within individual cells by cell walls. A visible collection of hyphae is called a fungal body, **mycelium,** or mycelial mat. The mycelium itself is colorless. Very commonly, colored spores are observed suspended above the mat and give the growth a fuzzy,

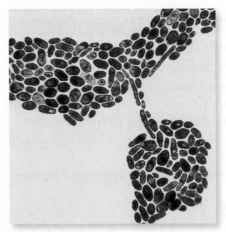

**Figure 6.2 *Saccharomyces cerevisiae.*** *S. cerevisiae* is commonly known as baker's yeast. It is a member of the phylum Ascomycota. Notice that the cells vary in size and shape and that the cells contain visible inclusions. (Magnification: 1,000×) ©Susan F. Finazzo

colored appearance. You have probably seen the bluish-green spores of *Penicillium*, the common bread mold. Some hyphae are modified and function to anchor the mycelium to its substrate. These anchoring hyphae are called rhizoids **(figure 6.3)**.

The fungi are not plants. They cannot perform photosynthesis. They are heterotrophs and must obtain their nutrients from the environment. They do this by secreting digestive enzymes into their surrounding environment. These enzymes break down or digest macromolecules. These smaller digested components can then diffuse more quickly back to, and be absorbed by, the hyphae. This ability to decompose complex macromolecules like lignin (plant cell wall component), cellulose (plant cell wall component), and chitin (insect skeletal material) makes the fungi important in recycling nutrients within ecosystems.

The fungi grow best under slightly acidic conditions (pH 4 to 6). They also tolerate high osmotic conditions better than most bacteria. They are common spoilers of acidic, salty, or sugary foods. If your grape jelly is fuzzy or slimy, you probably have a fungal contaminant!

Fungi reproduce both sexually and asexually **(figures 6.4 and 6.5)**. The fungi are classified into groups by their type of sexual spore, or fruiting body.

The zygomycetes (Zygomycota), or conjugation fungi, asexually produce spores at the tips of reproductive hyphae called sporangiophores. These hyphae lift the spores above the mycelial mat, where they can be carried away by the wind. The fruiting body itself is called a sporangium **(figures 6.4a and 6.5)**. Each little dot shown in the sporangium is an individual spore that can grow into a new organism. The sexual spore of the Zygomycota is called the zygospore **(figures 6.4b and 6.5)**. The zygospore forms when the haploid gamete of a + strand fuses with the haploid gamete of a − strand. A single offspring results from the zygospore. *Rhizopus* is a zygomycete.

It is sometimes known as "fruit mold" due to its common occurence on strawberries and tomatoes. Other members of the Zygomycota cause rare but potentially fatal infections.

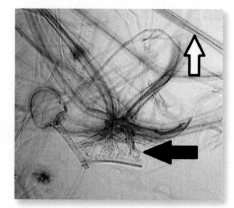

**Figure 6.3 Rhizoids and hypha.** The white arrow is pointing to a fungal filament or hypha. The black arrow is pointing to rhizoids, which are rootlike structures that anchor the mycelium. (Magnification: 400×)
©Susan F. Finazzo

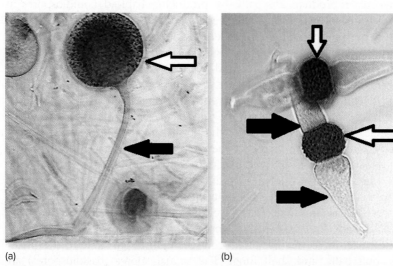

(a)                    (b)

**Figure 6.4 Asexual and sexual reproduction in members of phylum Zygomycota.** The image on the left **(a)** shows the asexual reproductive structure (white-filled arrow) of *Rhizopus stolonifer*. This structure is the fruiting body, or sporangium. The small bodies within the sporangium are individual spores. The black-filled arrow is pointing at the sporangiophore, which supports the fruiting body. Zygospores are shown in the right image **(b)**. The white-filled arrows are pointing at the zygospores. The black arrows are pointing at hyphae of the opposite mating types (+ and −). (Magnification: 400×)
©Susan F. Finazzo

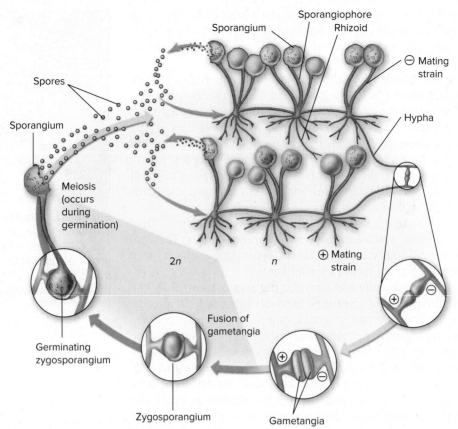

Spores

Sporangium

Sporangium

Sporangiophore

Rhizoid

⊖ Mating strain

Hypha

Meiosis (occurs during germination)

2n

n

⊕ Mating strain

Germinating zygosporangium

Fusion of gametangia

Zygosporangium

Gametangia

⊕

⊖

⊕

⊖

**Figure 6.5** Life cycle of *Rhizopus.*

There are approximately 30,000 species of the ascomycetes, or sac fungi. This group includes the yeasts and some mushrooms (morels, truffles). It also includes many economically important plant pathogens. The ascomycetes reproduce asexually by producing haploid conidiospores at the ends of reproductive hyphae called conidiophores. Yeast can reproduce asexually by budding. One cell undergoes mitosis to produce two nuclei. However, the new cell appears as a smaller attachment to the original cell. This smaller cell can separate or bud off of the original cell. Some species have buds that do not detach readily from each other, producing long strands of connected cells called **pseudohyphae.** Pseudohyphae production is a common presentation of *Candida albicans,* an opportunistically pathogenic yeast. *Candida* is part of the normal microbiota of many areas of the body and can cause disease of the oral cavity or vaginal mucosa, or less commonly it can cause systemic infection. Sexual reproduction in this group occurs when haploid structures produced by the + and − strands unite to form an ascus. The spores are held within the ascus. The ascus matures and ruptures to release multiple offspring. *Saccharomyces, Penicillium* **(figure 6.6),** *Aspergillus,* and *Candida* are all examples of ascomycetes.

The basidiomycetes (Basidiomycota), or club fungi, include the mushrooms, puff balls, and shelf fungi. The sexual spores (basidiospores) are formed by meiosis on special, club-shaped structures located on the gills on the underside of the mushroom cap. One disease-causing member of this group is *Cryptococcus,* which can cause meningitis, especially in patients immunocomprimised by conditions such as HIV infection or organ transplantation.

The last group is the deuteromycetes (Deuteromycota), or the Fungi Imperfecti. This group includes all of the fungi whose sexual spore has not been identified (sometimes known as anamorphs). If only the + strand

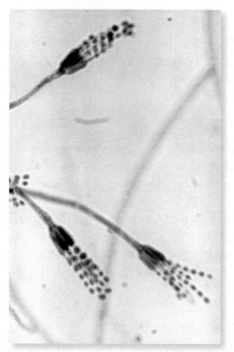

**Figure 6.6 Asexual reproductive structures of _Penicillium._** The asexual spores of _Penicillium,_ conidiospores, are the small, blue, round to oval structures. They extend from the conidiophore, which is a specialized reproductive hypha that supports the conidiospores and elevates the spores above the vegetative mycelium. Source: CDC/Lucille Georg

is growing in pure culture, the sexual spore will never form and the fungus cannot be classified. Today, mycologists are using gene sequencing to characterize and correctly classify members of the Deuteromycota. As organisms are analyzed, they are moved to the correct phylum. The number of genera included in the Deuteromycota is getting smaller and smaller as molecular geneticists learn about their hidden ancestry. At some point in the future, this group will no longer exist.

## ✓ Tips for Success

- Slide mounts that include agar blocks are very thick and can be easily cracked. Be careful when rotating objectives over the slide.
- You can view most slides at low power or high dry. You will need to use oil immersion for the yeast slide.
- Locate the stained specimen on your slide. **Some** slides have small blocks of agar on them. View around the edges of the colored agar block.
- You will need to adjust the light levels appropriately and possibly lower the microscope condenser for optimal viewing.
- **Do not open** any prepared Petri dish containing live specimens. Spores spread easily, and by opening the lid you risk contaminating your working area.

## Organisms (Prepared Slides)

- _Rhizopus_ zygospores
- _Penicillium_ or _Aspergillus_
- _Saccharomyces_ or _Candida_

## Materials

- Sabouraud plates on which sporulating fungi are growing
- Microscope
- Dissecting microscopes
- Lens cleaner and paper

## Procedure

1. Obtain a prepared slide from the supply table.
2. View and record your observations in the lab report. Use oil immersion for viewing yeast. For viewing the rest of the prepared slides, 100× or 400× total magnification should be adequate.
3. When you have viewed all of the available slides, clean and return your microscope to the cabinet.
4. Observe Sabouraud plates using the dissecting microscope or stereomicroscope. Do not open the Petri dishes. Record your results in your lab report.

## Results and Interpretation

As you look at the prepared slides of the molds, look for the features they all have in common and for the differences in their fruiting bodies. If you were looking at an unknown organism, how would you know it was a mold? While observing the yeast slides, contrast them with bacteria you have observed in other exercises.

# The Fungi

## Your Results and Observations

In the circles below, draw your specimens; indicate the magnification. Label any distinctive structures (conidio-spore, rhizoid, etc.). These circles can be used for slides or wet mounts of organisms not listed in the laboratory manual. You will need to fill in the organisms' names. Alternatively, you can use these circles to depict different magnifications of the same slide. For example, you can draw *Rhizopus* as it appears at 40× in the left circle and then use the circle to the right to draw *Rhizopus* at 100×. Your instructor will provide guidance.

*Rhizopus* Magnification: _____

Organism: _____

Magnification: _____

*Aspergillus* Magnification: _____

Organism: _____

Magnification: _____

*Saccharomyces* Magnification: _____

Organism: _____

Magnification: _____

## Interpretation and Questions

1. Fungal infections like athlete's foot are challenging to treat and often require long-term courses of antifungal drugs. Why do you think fungal infections are so difficult to treat? (*Hint:* Think about the mode of action of antibiotics and similarities or differences between fungi and the human host.)

_____

_____

_____

_____

2. Students commonly confuse *Saccharomyces cerevisiae* and *Staphylococcus aureus* when viewed on a microscope slide. How could you microscopically differentiate *S. cerevisiae* from *S. aureus*?

_____

_____

_____

_____

3. Describe how the asexual reproductive structures of *Rhizopus* and *Penicillium* (or *Aspergillus*) differ.

_____

_____

_____

_____

EXERCISE

7

# Bacterial Motility: Hanging Drop Method

## CASE FILE

An 82-year-old male has been in the intensive care unit at a Long Island hospital for 9 days following surgery for a subdural hematoma caused by a fall at home. He has recently begun spiking a fever of 102.7°F (39.3°C) and exhibiting breathing difficulty. Sputum and blood cultures sent to the lab return the next day positive for a gram-negative bacillus. Urine cultures are negative. The attending physician suspects that the causative bacterium may be a strain of either *Klebsiella* or *Enterobacter*. Treatment with intravenous (IV) antibiotics was initiated pending further testing to definitively identify the bacteria.

Source: CDC/Dr. Erskine Palmer & Byron Skinner

A variety of bacteria cause nosocomial (hospital-acquired) pneumonias. The most common causes include *Pseudomonas, Staphylococcus aureus,* and coliform bacteria such as *Escherichia coli, Klebsiella,* and *Enterobacter.* Members of two of these genera, *Klebsiella* and *Enterobacter,* give many identical results on biochemical tests (such as the ones you will be doing later in the term). One way of distinguishing them from each other is motility testing, since *Enterobacter* is motile and *Klebsiella* is nonmotile.

### LEARNING OUTCOMES

At the completion of this exercise, students should be able to
- perform and describe the hanging drop method for determining bacterial motility.
- be able to distinguish true motility from Brownian movement.

## Background

For many pathogenic bacteria, the ability to move toward targets inside the host is essential to their ability to cause disease. Most motile bacteria use **flagella** to swim. Flagellar movement often causes a "run-and-tumble" phenomenon where the bacteria change directions fairly frequently **(figure 7.1).** One way to determine if bacteria are motile is to observe them microscopically. In the hanging drop method, a drop of fluid containing bacteria hangs under a coverslip in the well of a depression slide. The

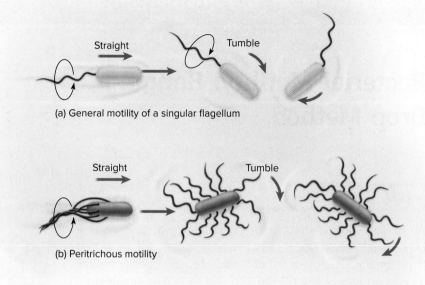

(a) General motility of a singular flagellum

Straight

Tumble

(b) Peritrichous motility

Straight

Tumble

**Figure 7.1** "Run-and-tumble" movement in bacteria with a single flagellum (a) and multiple flagella (b).

bacteria swimming (or not) in the drop are observed at 400× or 1,000×. Other methods that can be used include the wet mount and the use of soft agar (you will use soft/SIM agar in a later exercise).

There are two types of movement you may see: true motility and Brownian movement. **Brownian movement** is typically caused when water molecules collide with the cells and make them move. They get bounced around like a pinball. The cells oscillate and appear to shake in place (they don't move very far in any particular direction). This is not real motility.

## ✅ Tips for Success

- Because of the lack of contrast with their background, focusing on clear, unstained bacteria in water is a bit of a challenge. Try focusing on the edge of the water drop or drops first, and then you will have the bacteria in focus too.
- Controlling the amount of light using the lever on the diaphragm will be important. You will need to use less light than you would to look at a stained slide.
- Watch the slide for a while. Even when looking at truly motile bacteria, not all of them will be moving all the time.

## Organisms (on Slant or Plate Cultures—Handle These Pathogens with Caution)

- *Pseudomonas aeruginosa*
- *Micrococcus luteus*

## Materials (per Team)

- Depression slide
- Coverslips
- Vaseline or other petroleum jelly
- Toothpicks
- Disinfectant
- Water bottle
- Plain slide (optional)

## Procedure

1. We will be doing a hanging drop preparation today–you may use this or a regular microscope slide (wet mount) later in the term as part of the process of identifying your bacterial unknown.

2. Make a slide for each organism (if you have only one slide available, do the organisms one at a time).

3. Use your loop to put a small drop of water on a coverslip (steps 1 and 2 in **figure 7.2**).

4. Aseptically remove a small amount of the culture from the slant or plate. Mix the bacteria into the drop on the coverslip (step 3 in figure 7.2).

5. Using a toothpick, put a small amount of petroleum jelly on the edges of the coverslip (step 4 in figure 7.2).

6. Turning it upside down, place the depression slide down on the coverslip, so that the well of the slide is centered over the drop of water **(figure 7.3)**.

7. Now you can turn the slide so it is upright **(figure 7.4)**. Locate the water drop under the coverslip **(figure 7.5)** so you can position the objective lens to look at the bacteria with your microscope.

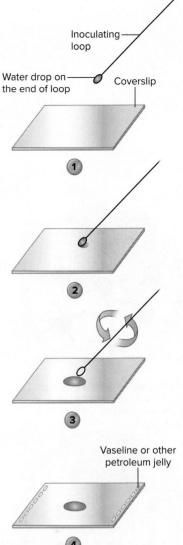

**Figure 7.2** Adding bacteria to the water on the coverslip and placement of the petroleum jelly.

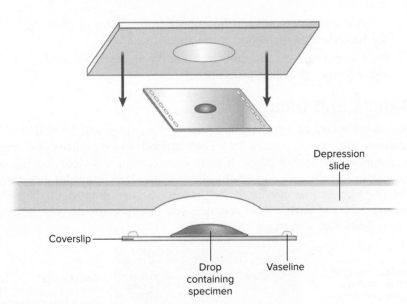

**Figure 7.3** Adhering the coverslip to the slide.

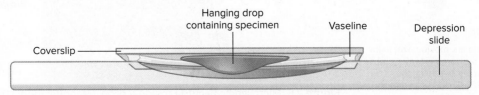

**Figure 7.4** The depression slide and hanging drop after being turned over.

55

**Figure 7.5** Depression slide with coverslip and water drop. ©Steven D. Obenauf

8. You will find that focusing on these unstained bacteria in water drops will be challenging. Close the iris diaphragm, so that very little light is passing through the specimen, and try to focus on the edge of the drops **(figures** 7.5 **and 7.6)** before looking at the bacteria. Use 400× or 1,000× total magnification.

9. Take some time and watch the bacteria for a while to find out if you are seeing true motility or Brownian movement **(figure 7.7)**. Remember, even motile bacteria are not moving all the time!

10. Record your results.

11. Clean off the depression slides with antibacterial soap when you are finished or place them in a beaker of disinfectant. **Do not discard them.** Put them back where they were.

12. Discard used coverslips and toothpicks in a sharps container.

## Wet mount (optional)

1. Use your loop to put a small drop of water on a plain microscope slide.

2. Remove a small amount of the culture from a slant or plate. Mix the bacteria into the drop on the slide.

3. Place a coverslip on the drop and observe as described in steps 8 and 9 of the preceding procedure.

## Results and Interpretation

Due to their lack of color, the bacteria in your drop will be difficult to focus on and observe (figure 7.6). Look around–you may have one drop or a number of smaller drops. If you observe many cells not moving or moving a short distance and a few cells moving a significant distance, your interpretation is *true motility*. If you observe a large number of the cells moving back and forth a short distance, your interpretation is *false motility* due to Brownian movement (figure 7.7).

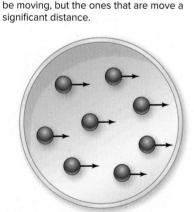

In Brownian motion, all of the cells will be "jiggling"— moving a little bit but in no particular direction.

In true motility, many of the cells may not be moving, but the ones that are move a significant distance.

If all of your bacteria are moving a lot in the same direction, make sure the water drop is not moving and simply sweeping the bacteria along with it. This is false motility.

**Figure 7.7** True and false motility.

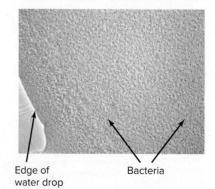

**Result:** Some cells rapidly moving a significant distance

**Interpretation:** True motility

**Result:** Large numbers of cells moving back-and-forth

**Interpretation:** False motility due to Brownian motion

Edge of water drop          Bacteria

**Figure 7.6** Microscopic bacteria inside a drop. (Magnification: 1,000×)
©Steven D. Obenauf

Name _____

Date _____

# Bacterial Motility: Hanging Drop Method

## Your Results and Observations

Record your results.

| Organism | Describe what the movement looks like. | True motility? (yes/no) |
|---|---|---|
| *Micrococcus luteus* | | |
| *Pseudomonas aeruginosa* | | |

## Interpretation and Questions

1. What is the most important thing to look for to determine whether you are seeing true motility or Brownian movement?

   _____

   _____

   _____

   _____

   _____

   _____

   _____

2. Why is it hard to see the cells in a hanging drop slide (your own experience doing this exercise or information in the "Results and Interpretation" section might help here)?

   _____

   _____

   _____

   _____

   _____

   _____

   _____

# Introduction to Staining

## Why Is Staining Important?

Visualizing cells is often the first step in the identification of bacteria. Bacteria, like most cells, are filled with clear, uncolored cytoplasm. Cells lack contrast to the surrounding medium and in an unstained state are difficult to see. Compounding this situation is the fact that bacteria are very, very small, on the order of micrometers. Small size and the lack of contrast to the background are two important factors affecting our ability to see and characterize bacterial cells.

## How Do Stains Work?

Staining techniques have been developed over the years that allow the experimenter to visualize the cell or cell structures. These staining techniques are based on the chemical characteristics of both the staining chemicals and the cells themselves. Stains are composed of a chromophore and a solvent. Basic stains have a positively charged chromophore. Since opposite electrical charges are attracted to each other, basic stains are attracted, and bind, to negatively charged moieties. Bacterial cells are typically negatively charged, so basic stains dye the cell. Acidic stains have a negatively charged chromophore. The charge of an acidic stain is repelled by the negative charge associated with a bacterial cell; therefore, acidic stains are repelled by the cell. Acidic stains dye the background and leave the cells uncolored.

## What Types of Staining Techniques Can Be Done?

There are three broad categories of staining techniques: simple staining, differential staining, and structural staining. Simple staining techniques use only a single dye or stain. The amount of information we can learn about the bacterium is limited to cell shape or morphology (coccus, bacillus) and cell arrangement (single cell, diplococci, chains, etc.). Differential staining uses two or more stains and allows for the identification of specific groups of organisms. The Gram stain and acid-fast stain are two examples of differential stains that are discussed more thoroughly in other exercises. Structural stains can also use multiple dyes–however, they are designed to stain a particular structural component of a cell. Examples of structural stains include the endospore stain and the flagella stain.

# Simple Staining and Smear Preparation

## CASE FILE

A 6-year-old boy complained to the school nurse that his ear was itchy and painful. The child did not have a fever. He was sent home and subsequently seen by his family physician. The external ear canal had a slight watery discharge, and visual examination revealed an inflamed and erythematous canal. The child complained of pain when the pinna was manipulated. The tympanic membrane appeared normal. The parent stated that since the onset of hot weather the child had spent every afternoon in the community pool. The doctor suspected acute otitis externa, commonly known as "swimmer's ear," and prescribed a topical combination drop composed of polymyxin B, neomycin, and hydrocortisone. She also suggested an oral analgesic for pain. Polymyxin B and neomycin are active against bacteria. Hydrocortisone is an anti-inflammatory administered to decrease edema and inflammation. The doctor recommended that the child stay out of the pool for 1 week until the infection cleared and that his ears be protected while bathing. Additionally, the doctor recommended as prophylactic measures wearing earplugs to the pool or that his parent rinse the child's ears with a solution of one part white vinegar and one part rubbing alcohol after swimming.

©Design Pics/Darren Greenwood RF

Swimmer's ear can be the result of trauma to the ear canal (e.g., from cotton swabs), blockage of ceruminous glands, or an increase in humidity in the canal, just to name a few potential causes. *Pseudomonas* spp. (gram-negative bacillus) and *Staphylococcus aureus* (gram-positive coccus) are the most common causative agents of swimmer's ear, although other bacteria and fungi can also be involved.

### LEARNING OUTCOMES

At the completion of this exercise, students should be able to

- successfully make a bacterial smear.
- identify the cellular morphology and arrangement of a bacterial culture.
- explain the purpose of heat-fixing and staining procedures.
- describe the general chemical basis for microbiological staining.

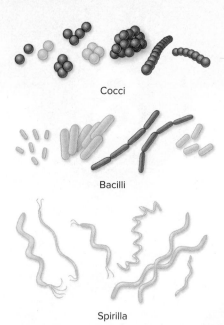

Cocci

Bacilli

Spirilla

**Figure 8.1 Bacterial morphology and arrangement.** Bacterial cells typically have one of three shapes: spherical (coccus), oblong (bacillus), or curved (spirillum or vibrio). Depending on the plane of binary fission, cells form chains, clusters with various geometry, or single cells.

# Background

Simple stains use only one stain to complete the procedure. Simple staining techniques can use basic stains, like safranin or methylene blue, which stains all of the bacterial cells on your slide one color. Procedures that stain cells are called direct stains. If an acidic stain is used, the background, not the bacterial cell, is colored. This type of simple stain where the background and not the cell is stained, is called negative staining. Negative staining is addressed in another exercise.

Simple staining is useful for observing the morphology or shape and the arrangement of cells. The three most common shapes of bacteria are the bacillus (oblong, rod), coccus (sphere), and spirillum (comma-shaped, spiral). Arrangement is based on the plane of cellular reproduction; individual cells are arranged singly, in groups of two or four, in chains or in clusters. The prefix used to describe the arrangement where two cells remain attached after fission is *diplo-*, so spherical cells that are observed in packets of two are called diplococci. When cells form long chains, the arrangement is called *strepto-*. A chain of rod-shaped cells would morphologically be described as streptobacilli. Grapelike clusters of cells exhibit the *staphylo-* arrangement. Grapelike clusters of cocci are called staphylococci. Some examples are shown in **figure 8.1.**

In order to actually stain bacteria, you must first make a smear. A smear is a thin film of bacteria applied to a slide. A good smear will have enough cells present to allow the observer to see cultural characteristics. A thick smear has too many cells; this can lead to clumping and will make observing cultural characteristics more difficult. Making a good smear and properly heat-fixing the smear are essential for achieving good results, not only in simple direct staining but in most other staining techniques.

During this class, you will work with cultures growing in liquid media, like broths, or on solid media, like slants and plates. A loopful of sample from a broth culture is more dilute and contains fewer cells than a loopful of inoculum removed from a solid agar plate surface. If your inoculum is growing on a slant or plate, you must first add a loopful of water to the slide to dilute the sample and to make it easier to spread the cells into a thin smear. Once you have added the drop of water, you will use your loop or needle to aseptically acquire a small sample of your culture. Mix the bacteria into the drop and spread the drop out to produce a thin film. In preparing the smear from a broth culture, you add a loopful of culture directly to the slide without diluting the sample further. You then use the loop to spread the drop into a thin film.

## ✔ Tips for Success

- When making a smear from a culture growing on a solid medium, add only one loopful of water to the slide to make the smear. Slides must air-dry before being heat-fixed. If you add too much water, you will need to wait a longer time for the water to evaporate.

- Add only a small amount of inoculum to the water drop. One loopful of a broth or a comma-size inoculum from a solid medium will provide more than enough cells to make the slide. Too many bacteria will make the smear too thick and difficult to view. The smear will stain unevenly and will not heat-fix properly.

- Spread the smear adequately. The smear should be a very thin film, not big chunks of bacteria.

- Do not forget to heat-fix your slide. Bacteria will wash off slides that are not heat-fixed.

- Make sure to place your slide on the microscope with the **smear side up.** If the slide is placed on the microscope upside down, the bacteria will not be visible because bacteria are too small for the plane of focus.
- Use a wax pencil or indelible marker to outline the smear on the bottom of the slide.

## Smear Preparation and Simple Stain Procedure

In this activity, you and your partner will each practice the basic techniques required for most staining activities by properly making a smear from a bacterial culture and completing a simple stain. Mastering the techniques needed to perform a simple stain will facilitate your mastery of the Gram, endospore, and acid-fast stains. You will use your slide to identify bacterial morphology and cellular arrangement.

## Organisms (Slants or Plates)

- *Bacillus cereus*
- *Staphylococcus aureus*

## Materials (per Team of Two)

- Slides
- Staining tray
- Slide holder or clothespin
- Bunsen burner and striker
- Methylene blue
- Safranin
- Water bottle
- Bibulous paper or paper towel
- Loop

## Procedure

It is important that you observe examples of both bacterial cultures and both stain colors. Each student will prepare one slide. Within your team of two, you should have two slides. One student will use *Bacillus cereus* as the specimen, and the other student will use *Staphylococcus aureus* when preparing the smear. One student will use methylene blue, and the other will use safranin.

1. Obtain a clean, dry slide. If water beads up on the slide, you may need to wash the slide with soap and water.
2. If you are using a broth culture, add one loopful of culture to the slide. If you are using a slant or plate, add one loopful of water to the slide first and then aseptically add a comma-size amount of bacteria to the water on the slide. (*Remember:* One partner is using *Bacillus*, and the other is using *Staphylococcus*.) Do not add too much water. You need only enough water to spread your inoculum.
3. Mix the bacteria into the drop of water and spread the drop until it is nickel- or quarter-size. Allow the slide to air-dry. (*Hint:* To more easily locate the smear after staining, use the indelible marker or wax pencil to draw a circle around the smear on the **BOTTOM** of the slide.)
4. Once the slide has air-dried (no visible water left), heat-fix the slide. Hold the slide using either the clothespin or slide holder. Then pass the slide through the flame of the Bunsen burner, about 3 to 4 inches above the burner. **DO NOT** hold the slide directly in the flame for more than a second or two. If you leave the slide in the flame, you will burn off the bacteria, and the slide may break. Repeat this procedure four or five times. The smear should now be heat-fixed.
5. Place the slide on the staining tray.
6. Cover the smear with either safranin or methylene blue. (One partner should use safranin; the other partner should use methylene blue.)

©Steven D. Obenauf

7. Allow the smear to stain. If you are using safranin, allow the smear to stain for about 1 to 2 minutes. If you are using methylene blue, allow the smear to stain for 4 to 5 minutes.

8. At the end of the staining period, use the slide holder to tip the slide, so that the excess stain runs off of the slide and into the staining tray.

9. Use your water bottle to **GENTLY** rinse the slide for about 10 seconds. Squirt the water onto the slide away from the smear and slowly rinse the entire slide. Do not aim a high-intensity jet of water directly at your smear–it will wash off.

10. Blot your slide dry by inserting the slide between the sheets of bibulous paper and gently sliding your hand across the book of bibulous paper. Do not press down on your slide–it will break.

11. Clean the staining tray by disposing of the used stain in the stain discard container. Rinse and dry the staining tray. Return all materials to their proper location.

12. Slant stock cultures will be used in the next class. Return slants to the incubator.

13. Remove the slide from the bibulous paper and view using oil immersion. Remember, start focusing on your slide using the 4× objective and then move to progressively higher magnifications. You do **not** need to remove the stained pages of bibulous paper.

14. Record your observations of your slide in your lab report. View your partner's slide and record that information in your lab report also. Remember to include the total magnification used with your drawing, the stain used, and the name of the culture. Use a colored pencil for drawing the cells.

15. If you would like to keep your slide to view at another time, gently blot the oil from the slide with a paper towel. Do not rub the slide or use lens cleaner on the slide. Store your slide in your group's labeled slide box. Otherwise, discard your slide in the sharps container.

## Results and Interpretation

Staining techniques are essential tools in the microbiologist's toolbox. They allow the scientist to visualize cells and–depending on the staining technique–identify or classify bacteria. Simple staining techniques utilize a single stain and reveal the shape (morphology) and the arrangement of cells in the culture **(figure 8.2).**

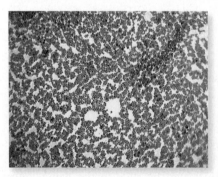

(a)

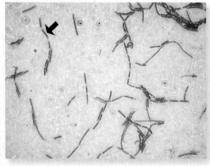

(b)

**Figure 8.2 Safranin and methylene blue simple stains.** Image **(a)** is a safranin-stained slide of *S. aureus.* Can you determine the morphology and arrangement shown here? Image **(b)** is from a slide of a *B. cereus* culture stained using methylene blue. Can you determine the morphology and arrangement shown here? ©Susan F. Finazzo

Name _____

Date _____

# Simple Staining and Smear Preparation

## Your Results and Observations

In the circles below, draw what you observed on your slide and on your lab partner's slide. Indicate the magnification and stain used in preparing these slides.

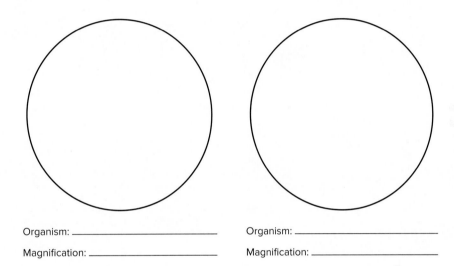

Organism: _____

Magnification: _____

Organism: _____

Magnification: _____

## Interpretation and Questions

**1.** You forgot to heat-fix your slide. What would you expect to see?

_____

_____

_____

_____

_____

**2.** How would you describe the morphology and arrangement of the cells in your stained preparations?

_____

_____

_____

_____

**3.** You are interning in the doctor's office described in the case file. You use some of the ear exudate to make a slide, which you then heat-fix and stain. Under the microscope, you see large clusters of spherical cells. Which organism are you probably seeing: *Staphylococcus* or *Pseudomonas*? Explain.

_____

_____

_____

_____

_____

# Negative Stain

## CASE FILE

A 32-year-old male was seen in an emergency triage clinic
following an earthquake and landslide. He presented with a
rash on the palms of his hands and the bottoms of his feet.
The rash did not itch and appeared as rough, reddish-brown
spots. Additionally, the patient had a slightly elevated temperature and swollen lymph nodes. A whole-body exam was
performed. The doctor noted a nearly healed chancre on the
patient's genitalia. When questioned, the patient reported the
chancre had appeared 4 weeks earlier but was painless and therefore
he had not visited the local clinic. The doctor sampled the chancre
area, and a negative stain revealed the presence of spirochetes. The
patient was treated with intramuscular penicillin and released.

Source: Centers for Disease
Control

## LEARNING OUTCOMES

At the completion of this exercise, students should be able to
- properly prepare a negatively stained slide.
- explain the basis for and advantages of negative staining.

## Background

The negative stain is an example of an indirect simple stain. The dyes used
in negative staining have an acidic chromogen (negatively charged chromophore) that is repelled by the bacterial cell. As a result, the cell remains
uncolored and the background is stained. Nigrosin and eosin are examples of
negative stains. Heat-fixing smears can distort cell size and cell shape. Since
negative stain slides are not heat-fixed, cell shape and size are not distorted
and can be accurately observed. Negative staining is particularly appropriate
for visualizing delicate structures that are distorted by heat like capsules and
cells like the spirilla and spirochetes such as *Treponema pallidum*, the cause of
syphilis. Negative staining is also useful for visualizing bacteria that are not
stained easily using traditional stains, like the Gram stain. An example of such
a bacterium is *Treponema*, the cause of syphilis, the disease in the case file.

## ✔ Tips for Success

- Make the smear with one continuous, fluid motion. There will be
  thick and thin areas of the slide.
- Place the drop of stain near one end of the slide, not in the center.

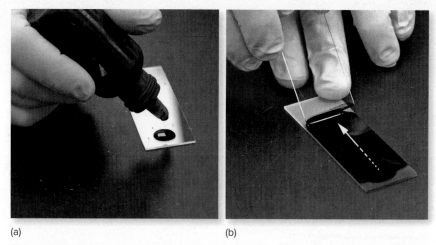

(a)                                    (b)

**Figure 9.1 Creating the negative stain smear. (a)** Apply a drop of nigrosin near the end of a clean slide. Mix bacteria into the nigrosin drop. **(b)** Use the second clean slide to spread the nigrosin across the slide. Some areas of stain on the slide will be thinner than others. ©Steven D. Obenauf

## Organisms

- *Bacillus* (slants or plates)
- *Klebsiella* (milk broth)

## Materials

- 2 clean slides
- Loop
- Staining tray
- Nigrosin
- Coverslip
- Bunsen burner and striker

## Procedure

1. Place your slide on the staining tray or benchtop.
2. Add one drop of nigrosin to one end of the slide (not in the middle) **(figure 9.1a)**.
3. Aseptically add one loopful of *Klebsiella* to the drop of nigrosin.
4. Flame your loop.
5. Aseptically add a small amount of *Bacillus* to the same drop of nigrosin in order to create a mixture of the two organisms.
6. Pick up the second clean slide. You will use this slide to spread the smear. Hold the clean slide at about a 30-degree angle to the slide.
7. Start on the end of the slide away from the dye. Slide the clean slide toward the nigrosin drop until the dye runs along, the second slide edge and across the width of the slide.
8. In one single motion, pull the second slide quickly down the length of the slide **(figure 9.1b)**. This should result in a long, dark smear with varying thicknesses along the entire length of the slide.
9. Allow the slide to air-dry.
10. When the slide has air-dried, cover the smear with a coverslip and view under oil immersion.

## Results and Interpretation

The stain stains the background, not the cell. Cells will be unstained and appear bright white against a dark background **(figure 9.2)**. You may have to tell the cells apart from debris or bubbles on your slide.

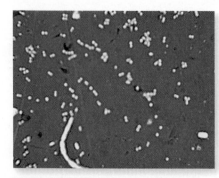

**Figure 9.2 Negative stain.** A drop of nigrosin inoculated with *Bacillus* and *Klebsiella* was used to produce this slide. The cells appear as bright white shapes against a dark background. ©Susan F. Finazzo

## Lab Report 9

Name _____

Date _____

# Negative Stain

## Your Results and Observations

In the circle below, draw (or photograph with your mobile device) what you observed on your slide. Label appropriately. Indicate the magnification.

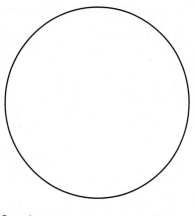

Organisms: _____

Magnification: _____

## Interpretation and Questions

**1.** Why didn't you heat-fix this slide?

_____

_____

_____

_____

**2.** How would you describe the morphology and arrangement of the cells in your stained preparations (you may want to refer to the "Background" section of exercise 8 to assist you here)?

_____

_____

_____

_____

**3.** What morphological characteristics would you use to distinguish between *Klebsiella* and *Bacillus* on the slide?

_____

_____

_____

**4.** Explain why the cells remain unstained and the background is colored.

_____

_____

_____

_____

# Capsule Stain

## CASE FILE

The family of a man asked the police department to check
on their 29-year-old relative, whom they had not been able to
contact. When the police arrived, they found the man uncon-
scious on the bed. Vomit and feces were evident on the
man's clothing and bedding. His airway was partially occluded
with vomitus. A policeman repositioned the man on his side to
facilitate breathing and then left the room. Paramedics arrived,
measured the patient's vital signs, suctioned his airway,
administered oxygen, and then transported him to the
emergency room.

Source: Image courtesy of the
Centers for Disease Control and
Prevention/Joe Miller

In the emergency room and subsequently in the ICU, the patient
had an endotracheal tube inserted and had blood and cerebrospinal
fluid (CSF) drawn for culture and for staining (Gram, capsule).
Meningococcal infection was suspected. Treatment was initiated with
piperacillin-tazobactam, levofloxacin, ceftriaxone, and vancomycin.

Microscopic examination (as you will do today) of Gram and
capsule stains of the CSF indicated the presence of gram-negative
encapsulated diplococci. *Neisseria meningitidis* was isolated from
both blood and CSF cultures. The case was reported to the local
health authority. The patient was hospitalized for 3 weeks and then
released.

One day after admission of this patient, a 30-year-old policeman
that had attended the ill patient complained of a sore throat, nausea,
muscle pain, fever, and vomiting. The officer visited his primary
care physician and was sent directly to the emergency room. Blood
and CSF collected from the officer indicated the presence of
gram-negative diplococci. He was hospitalized, treated, and released
5 days later.

*Neisseria meningitidis* is contagious. Of the 10 health care and
emergency personnel that worked within 3 feet of this patient,
three contracted meningococcal disease. It is the leading cause of bacterial
meningitis. Although disease fatality is less than 15%, survivors of this
dangerous disease can suffer brain damage, loss of limbs or other
extremities, learning disabilities, and hearing loss.

## Background

Bacteria may produce an extracellular (outside the cell) structure called the **glycocalyx.** *Glycocalyx* literally means "sugar cover." If the glycocalyx is compact and firmly attached to the cell wall, it is called a capsule **(figure 10.1).** If the structure is less compact and loosely attached to the cell wall, it is called a slime layer. Many bacteria are thought to produce at least a minimal capsule, and some bacteria under favorable conditions (such as in the milk broth used in today's exercise) produce very thick capsules.

**Capsules** are composed of polysaccharides (sugars) and/or polypeptides (amino acids). Capsules can fulfill several functions: They serve as energy and water reserves, a means to evade the host's immune system, and an anchor or surface attachment mechanism. Bacteria produce thick capsules under favorable conditions. When nutrients are plentiful, bacteria store extra energy in the form of sugars and peptides in their capsules. When conditions are less favorable and nutrients are limiting, bacteria can then metabolize the molecules in their capsules as an energy reserve. Capsules make bacteria sticky. Sticky bacteria can adhere to surfaces. This is an advantage to bacteria like *Streptococcus mutans,* a common inhabitant of your mouth. *S. mutans* sticks to your teeth; food passes by frequently, nourishing the bacteria. Because of the capsule, the bacterium sticks to your teeth and doesn't get swallowed with the rest of your food. Unfortunately for us, *S. mutans* produces acids that weaken enamel and lead to the formation of **dental caries.** Capsules, because of their composition, also attract water. Capsules function as water-storing structures and assist bacteria in remaining hydrated. Finally, and of significant clinical importance, capsules enhance a bacterium's ability to evade phagocytosis and the body's immune system. Encapsulated forms of organisms are more

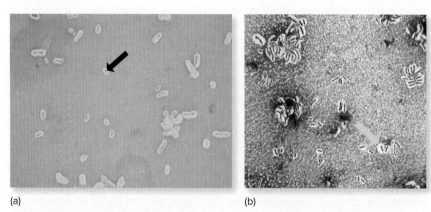

(a)                                         (b)

**Figure 10.1 Capsule stains of *Klebsiella pneumoniae.*** The slide shown in **(a)** was stained with safranin. The black arrow is pointing to a cell. The slide labeled **(b)** was stained with crystal violet. The yellow arrow is pointing to the capsule. Milk broth produces a background that varies from fine grain **(a)** to coarse clumps **(b).** ©Susan F. Finazzo

virulent because of this ability to "hide" from the immune system. *Streptococcus pneumoniae*, *Klebsiella*, *Neisseria*, and *Bacillus anthracis* all produce capsules that are important in their pathogenesis.

Capsules are delicate structures that are destroyed by heat-fixing or washed away during typical staining procedures such as a simple or Gram stain. Capsule staining techniques stain the bacterial cell and the background and leave the capsule intact and unstained.

## ✔ Tips for Success

- The smear for a capsule stain should be thicker than smears used for simple staining.
- If you are doing more than one lab exercise this period, make this smear first in order to give it time to air-dry.
- **DO NOT HEAT-FIX.**
- **DO NOT RINSE THE SLIDE WITH WATER.**
- When viewing the slide with oil immersion, look for areas of the slide where the background appears smooth and regular. These are often near the edge of the smear.

## Organisms

*Klebsiella pneumoniae* (milk broth culture)

## Materials

- Loop
- Clean slide
- Staining tray
- Safranin or crystal violet
- Copper sulfate
- Bunsen burner and striker

## Procedure

1. Place a clean, dry slide on the staining tray or benchtop.
2. Aseptically add one loopful of culture to the clean slide.
3. Spread the inoculum, so that your smear will be about the size of a nickel. This smear should be a little thicker than smears used in other staining techniques.
4. Air-dry. **DO NOT HEAT-FIX!**
5. Cover the smear with the primary stain. Use either crystal violet **or** safranin. Stain for 1 minute.
6. Tip the slide at a 45-degree angle and rinse the slide with copper sulfate for 20 to 30 seconds. **DO NOT RINSE WITH WATER.**
7. Blot the slide dry and view under oil immersion.

### Alternate procedure for making the smear

1. Place a clean, dry slide on the staining tray or benchtop.
2. Aseptically add one loopful of inoculum to one end of the slide.
3. Use another clean slide to spread the inoculum as done in exercise 9, "Negative Stain."
4. Air-dry. **DO NOT HEAT-FIX!**

5. Cover the smear with the primary stain. Use either crystal violet **or** safranin. Stain for 1 minute.

6. Tip the slide at a 45-degree angle and rinse the slide with copper sulfate for 20 to 30 seconds. **DO NOT RINSE WITH WATER.**

7. Blot the slide dry and view using the 100× oil immersion objective.

## Results and Interpretation

Some bacteria produce an extracellular glycocalyx. The ability to produce a glycocalyx, slime layer, or capsule can be advantageous to the organism and enhance its pathogenicity. Visualizing the capsule requires the use of a special staining procedure, the capsule stain. Stain is applied to a non-heat-fixed smear. The slide is then decolorized with copper sulfate. The stain will color the cell and the background (figure 10.1); the capsule will appear uncolored or light blue in color.

Name _____

Date _____

# Capsule Stain

## Your Results and Observations

In the circle below, draw what you observed on your slide. Indicate the magnification. Label a capsule and a cell.

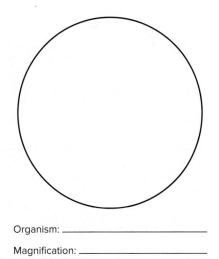

Organism: _____

Magnification: _____

## Interpretation and Questions

**1.** Why did you not heat-fix this slide?

_____

_____

_____

_____

**2.** Why is the presence of a capsule clinically significant (take some time and review the "Background" section thoroughly)?

_____

_____

_____

_____

# NOTES

# Gram Stain

## CASE FILE

A 52-year-old woman in Georgia visited her primary care physician complaining of pain, redness, light sensitivity, and swelling in her right eye. She reported that she had scratched her eye with a mascara applicator a day earlier. The doctor observed a corneal abrasion. He applied gentamicin ointment and patched the eye. Two days later, no improvement was observed and the abrasion had abscessed. The patient experienced impaired vision and was admitted for treatment. Gram staining of corneal scrapings from the abscess revealed the presence of gram-negative bacilli. The patient's mascara and corneal scrapings were cultured. Cultures from both samples revealed the presence of *Pseudomonas aeruginosa* sharing identical antibiotic sensitivity patterns.

Source: CDC/Janice Haney Carr

The infection was cleared up by subsequent antibiotic therapy. However, the patient experienced a dense inflammatory infiltrate of the cornea that was followed by neovascularization of the cornea. Her eyesight has not improved.

## LEARNING OUTCOMES

At the completion of this exercise, students should be able to
- properly prepare a Gram stain.
- interpret the Gram reaction observed on a slide.
- explain the chemical basis for the staining technique.
- describe the differences between the cell wall structures of gram-negative and gram-positive cells.

## Background

The Gram stain, developed by Hans Christian Gram in the late 1800s, is probably the most important and most frequently used staining technique in microbiology. It is usually the first step in the identification of an unknown organism isolated from a patient.

The Gram stain is a differential stain–that is, it differentiates by color two major groups of bacteria: the gram-positive and the gram-negative bacteria. The response (positive or negative) of cells to this staining

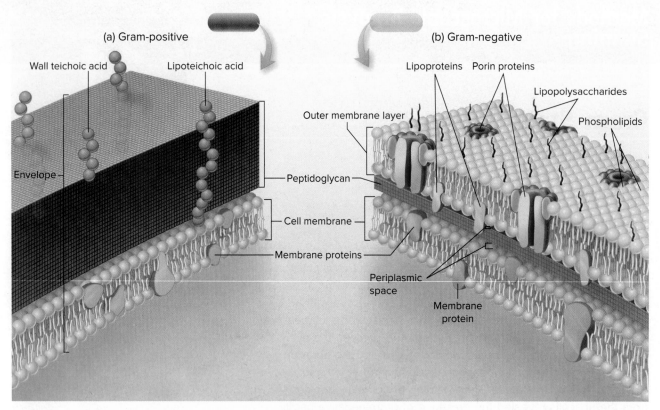

**(a) Gram-positive**

Wall teichoic acid

Lipoteichoic acid

Envelope

**(b) Gram-negative**

Lipoproteins   Porin proteins

Lipopolysaccharides

Phospholipids

Outer membrane layer

Peptidoglycan

Cell membrane

Membrane proteins

Periplasmic space

Membrane protein

**Figure 11.1  A comparison of the wall structure of gram-positive and gram-negative cells. (a)** An artist's rendering of a gram-positive cell wall. Notice the thick layer of peptidoglycan. **(b)** An artist's rendering of a gram-negative cell wall shows the thin layer of peptidoglycan and the inner and outer membranes.

technique is based on the composition of the cell wall **(figure 11.1).** The cell wall of gram-positive cells is made up of a very thick layer of peptidoglycan. Peptidoglycan is a long, fibrous network of NAM (*N*-acetyl muramic acid) and NAG (*N*-acetyl glucosamine) cross-linked by short peptide chains. The cell wall of gram-negative cells consists of a very thin layer of peptidoglycan surrounded by an outer membrane containing lipopolysaccharide (LPS).

The Gram stain technique **(figure 11.2)** involves the use of dyes/ stains, a mordant, and a decolorizer. The first step is to make a heat-fixed smear. Crystal violet, the **primary stain,** is applied to the heat-fixed smear for 1 minute. Crystal violet is a basic stain and stains all of the cells on the slide purple. The slide is rinsed with water and then the smear is covered with Gram's iodine. Gram's iodine is a **mordant.** Mordants fix color into the cell or intensify color. In this case, the iodine binds to RNA and crystal violet in the cell, creating a **very large** color complex. The next step is **decolorization** with acetone-alcohol (95% ethanol may be used instead). The slide is rinsed with acetone-alcohol. Acetone-alcohol is a solvent/denaturant. It creates large holes in the gram-negative cell's outer lipopolysaccharide cell wall and washes out the iodine–crystal violet complex. The thick peptidoglycan layer in gram-positive cells retains the iodine–crystal violet complex. After the decolorization step, gram-negative cells are colorless and gram-positive cells are purple. The final step is the application of the **counterstain,** safranin. Safranin is applied to the smear for 1 minute. The smear is then rinsed with water and blotted dry. Gram-positive cells will appear purple, and gram-negative cells will appear pink.

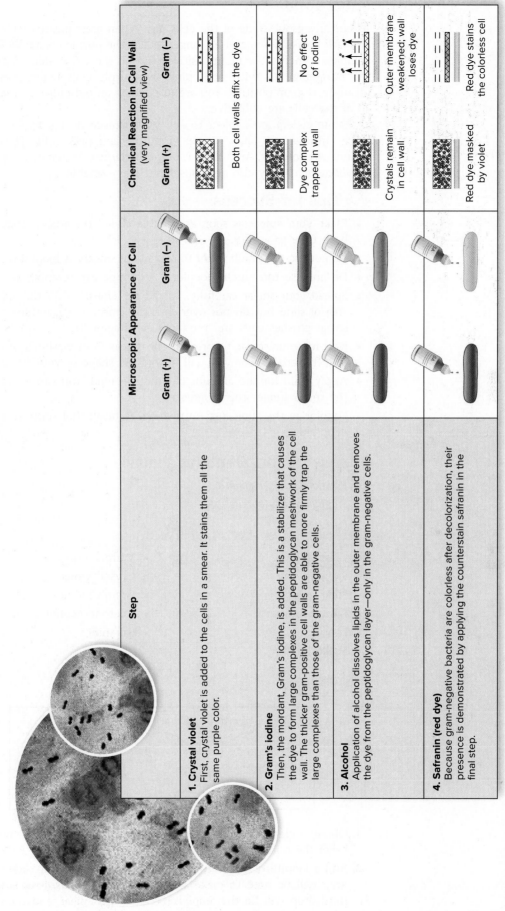

**Figure 11.2 Gram staining procedure.** The Gram staining procedure is shown here along with a representation of how gram-positive and gram-negative cells would appear at each step. ©McGraw-Hill Education/Neal R. Chamberlin, PhD

| Step | Microscopic Appearance of Cell | | Chemical Reaction in Cell Wall (very magnified view) | |
|---|---|---|---|---|
| | Gram (+) | Gram (−) | Gram (+) | Gram (−) |
| **1. Crystal violet** First, crystal violet is added to the cells in a smear. It stains them all the same purple color. | | | | Both cell walls affix the dye |
| **2. Gram's iodine** Then, the mordant, Gram's iodine, is added. This is a stabilizer that causes the dye to form large complexes in the peptidoglycan meshwork of the cell wall. The thicker gram-positive cell walls are able to more firmly trap the large complexes than those of the gram-negative cells. | | | Dye complex trapped in wall | No effect of iodine |
| **3. Alcohol** Application of alcohol dissolves lipids in the outer membrane and removes the dye from the peptidoglycan layer—only in the gram-negative cells. | | | Crystals remain in cell wall | Outer membrane weakened; wall loses dye |
| **4. Safranin (red dye)** Because gram-negative bacteria are colorless after decolorization, their presence is demonstrated by applying the counterstain safranin in the final step. | | | Red dye masked by violet | Red dye stains the colorless cell |

## Gram Variability

As cultures grow, some of the cells die. When gram-positive cells die, the cell wall breaks down and no longer retains the purple crystal violet-iodine complex. When older gram-positive cultures are Gram stained, the result can be ambiguous with both pink and purple cells being present. The organism is gram-positive; you are just seeing an old culture in which not all of the cells are still alive.

Some genera are subject to cell wall damage during fission. Stains of these cultures will show a mix of purple and pink cells. These genera include *Actinomyces*, *Arthrobacter*, *Corynebacterium*, *Mycobacterium*, and *Propionibacterium*. They also are considered gram-variable.

### ✔ Tips for Success

- Make your smear as soon as possible after lab starts, so that it will have time to air-dry.
- Do not use too much water to make your smears. A loopful is sufficient.
- Do not use too much inoculum. A comma-size scraping is sufficient.
- Spread your smear carefully. Spread the smear to fill the appropriate area of slide but do not overwork the smear. Overworking the smear crushes cells and produces cell fragments.
- Heat-fix your smear to prevent the smear from washing away. Don't heat-fix too long or you will distort the shape of your bacteria.
- Apply stain for the amount of time specified. You can always stain the smear longer than suggested.
- Decolorize the slide just until the acetone-alcohol starts to turn very light blue.

### Organisms (on Slants or Plates)

- *Pseudomonas aeruginosa*
- *Staphylococcus aureus*

### Materials (per Team of Two)

- Clean slide
- Staining tray
- Slide holder
- Bunsen burner
- Loop
- Striker
- Water bottle
- Crystal violet
- Gram's iodine
- Acetone-alcohol
- Safranin

### Procedure

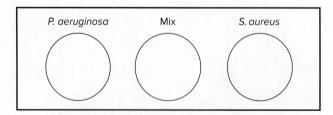

1. Obtain a clean slide. Use the preceding diagram to label your slide. Label the bottom of the slide.
2. Add a loopful of water to each of the circles on your slide. The first drop will be used to make a smear of *Pseudomonas aeruginosa*. The third drop will be the *Staphylococcus aureus* smear, and the center

drop will be a mixture of the two organisms. *Note:* If your organisms are in broth culture, do not add a drop of water to the slide. Add a loopful of inoculum to the appropriately labeled area of your slide.

3. Use the loop to aseptically remove a comma-size sample of *P. aeruginosa*. Mix the inoculum into the first drop and then the center drop. Spread the drops to create a thin smear.

4. After flaming your loop, aseptically remove a comma-size sample of *S. aureus*. Mix the *S. aureus* into the third drop of water and then into the center drop of liquid. Spread the drops to create a thin smear.

5. Allow the slide to air-dry.

6. Heat-fix the slide by moving the slide through the flame of the Bunsen burner several times. Place the heat-fixed slide on the staining tray.

7. Cover the smears with crystal violet, the primary stain. Stain for 1 minute.

8. After 1 minute, tip the slide and gently wash the slide with distilled water.

9. Cover the smears with Gram's iodine, the mordant, for 1 minute.

10. After 1 minute, tip the slide and drip the decolorizer acetone-alcohol down the slide. Decolorize for 15 to 20 seconds. The decolorizer dripping off of the slide will initially be very dark blue. After 15 to 20 seconds, the decolorizer color will be much lighter but still slightly blue.

11. Cover the smears with safranin, the counterstain, and stain for 1 minute.

12. Rinse the slide with distilled water and blot dry.

13. View using oil immersion (1,000×). Gram-positive cells will appear purple, and gram-negative cells will stain pink.

## Results and Interpretation

The Gram stain is probably the most important stain used in microbiology. It is a differential stain that distinguishes between bacterial cells based on their cell wall characteristics. After staining, gram-positive bacterial cells will appear purple, and gram-negative cells will appear light pink **(figure 11.3)**.

**Figure 11.3 Gram stain of a mixture.** This is a Gram stained slide of a mixture of *Pseudomonas aeruginosa* and *Staphylococcus aureus. S. aureus* is the gram-positive (purple) coccus. *P. aeruginosa* is the gram-negative (pink) bacillus at the arrow tip. ©Susan F. Finazzo

# NOTES

Name _____

Date _____

# Gram Stain

## Your Results and Observations

In the circle below, draw what you observed on your slide. Indicate the magnification.

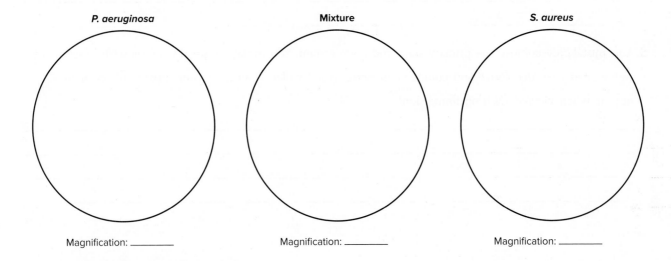

**P. aeruginosa**

Magnification: _____

**Mixture**

Magnification: _____

**S. aureus**

Magnification: _____

## Interpretation and Questions

1. What would you observe if you decolorized your slide too much? How would your cells appear (make sure you understand what is going on in each step of the Gram stain; that is key to answering this question as well as questions 3 and 5)?

_____

_____

_____

_____

2. How would you describe the morphology and arrangement of the cells in your stained preparations?

_____

_____

_____

_____

**3.** You are looking at the smear of the mixture. All of the cells, cocci and bacilli, appear deep purple. What could have gone wrong?

_____

_____

_____

_____

**4.** Explain the relationship between the observed Gram reaction and bacterial cell wall structure.

_____

_____

_____

_____

**5.** You mistakenly confuse the primary stain and counterstain. You initially stain the smear with safranin, add iodine, and then decolorize and counterstain with crystal violet. How does your mixed culture now appear when viewed with oil immersion?

_____

_____

_____

_____

# Acid-Fast Stain (Ziehl-Neelsen and Kinyoun Methods)

## CASE FILE

A 60-year-old male dairy farmer (Patient F) visited his primary care physician complaining of knee and hip pain. Two months earlier, the farmer had bruised his knee while moving hay bales. The bruising had been minor, and only in the last 3 weeks did he report experiencing joint pain and feeling ill. The patient also stated he had been losing weight, coughing, and feeling more fatigued than normal. The farmer lived alone in his Colorado community and had not had any long-term, sustained contact with anyone else. The doctor noted the knee joint was slightly swollen, but no other signs of inflammation were observed. Mobility in the joint appeared limited. The patient's lungs sounded congested. The doctor suspected the patient had tuberculous arthritis. The doctor ordered blood work, acid-fast staining of a sputum sample, a Mantoux tuberculin skin test, and a chest X ray. He asked the patient to return in 2 days.

Source: CDC/Dr. George P. Kubica

The patient's CBC (complete blood count) indicated an elevated monocyte count. The chest X ray showed several areas of patchy infiltrate with hazy borders. A 10-millimeter-wide area of induration (hardened, raised, and palpable) was present on the forearm where the Mantoux tuberculin skin test was done, indicating an active recent mycobacterial infection. Acid-fast staining (as you will be doing in this exercise) of his sputum revealed the presence of slender, acid-fast positive bacilli.

The patient was prescribed a 2-month regimen of isoniazid, rifampin, and pyrazinamide. This was followed by another 2-month prescription of isoniazid and rifampin. The infection resolved at the completion of this course of treatment.

## LEARNING OUTCOMES

At the completion of this exercise, students should be able to

- list the common genera of acid-fast bacteria and the clinically significant diseases they cause.
- correctly perform and interpret an acid-fast stain.
- explain the basis for the steps in the staining technique.

# Background

Acid-fast staining is a differential staining technique that differentiates between acid-fast and non-acid-fast bacteria. Acid-fast genera include the *Mycobacterium* and some members of the *Nocardia*. Pathogenic mycobacterial species include the causative agents of tuberculosis (TB) and leprosy. TB, in particular, is still a major health concern and cause of death throughout the world.

Acid-fast cells have a waxy cell wall. **Mycolic acid** in the cell walls of acid-fast bacteria makes these organisms resistant to desiccation (drying) and disinfectants. This waxy cell wall also makes it difficult to stain these organisms with traditional stains that are water soluble. Water-soluble stains are repelled by the cell wall in the same way that water beads up on a paraffin candle. There are two different means commonly used to stain acid-fast bacteria: (1) The Ziehl-Neelsen method uses a steaming water bath to facilitate staining of the cell, and (2) the Kinyoun method uses more concentrated reagents and does not require steam (cold method). The methodology for both techniques is provided here.

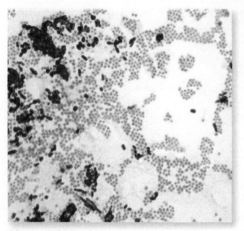

©Steven D. Obenauf

The Ziehl-Neelsen technique for acid-fast staining entails steaming heat-fixed slides over boiling water with the primary stain, carbol fuchsin. Heating "melts" or increases the fluidity of the waxy cell wall (mycolic acid) of acid-fast cells and permits the penetration of the primary stain. After staining, the slide is cooled. As the slide cools, the waxy cell wall firms up and loses its fluidity, trapping the stain in the cell. Smears are decolorized with acid alcohol. Cells that retain the primary stain after decolorization are called acid-fast–that is, the bright red fuchsia color remains fast in the cells. The slide is then counterstained with a secondary stain, methylene blue, to dye the non-acid-fast cells. Non-acid-fast cells will appear blue.

The steps in the Kinyoun acid-fast staining procedure are nearly the same as the Ziehl-Neelsen method. The Kinyoun method uses carbol fuchsin, acid alcohol, and methylene blue but does not require the slide to be steamed. The primary stain reagent contains higher concentrations of phenol and basic fuchsin than in the Ziehl-Neelsen reagent. Phenol is a lipid solvent. Increasing the concentration of phenol increases the solubility of the cell wall or "dissolves" the waxy cell wall and allows the basic fuchsin to penetrate the cells. The higher concentrations of phenol and basic fuchsin permit the staining of the cells at benchtop temperatures.

## ✓ Tips for Success

- Mycobacteria are very "sticky due to mycolic acid in their cell walls." Spread your smear thoroughly. You may still see clumps.
- Heat-fix the slide before staining.
- Place a small piece of paper towel over the smear to decrease spattering and to prevent crystals from forming directly on the smear when steaming. This step is not necessary if you are performing the Kinyoun staining technique.
- Slides **MUST** steam (Ziehl-Neelsen technique) with the carbol fuchsin stain for at least 5 minutes to dye the acid-fast cells.
- Do not let the slide dry out. Reapply stain as it evaporates.
- Cool slides before decolorizing.

## Organisms (Plate or Slant)

- *Mycobacterium smegmatis*
- *Staphylococcus aureus*
- Patient F culture

## Materials (per Team of Two or Four)

- Staining tray
- Slide holder
- Acid alcohol
- Methylene blue
- Forceps
- Water bottle
- Carbol fuchsin (Ziehl-Neelsen)
- Carbol fuchsin (Kinyoun)
- Hot plate
- Paper towel
- Beaker containing water
- Bunsen burner
- Loop

## Procedure (Ziehl-Neelsen)

1. Obtain a clean slide. If your hot plate is not on, turn it on.
2. Use your loop to add three drops of water to the slide to create three smears **(figure 12.1)**.
3. Aseptically remove a small, comma-size amount of the *S. aureus* culture and mix it into the first water drop and the center water drop. Spread the drops to create a thin smear.
4. Aseptically remove a small, comma-size amount of the *Mycobacterium* or Patient F culture and mix it into the third and the center water drops. Spread the drops to create a thin smear.
5. Allow the smears to air-dry.
6. Heat-fix the smear.
7. Place the heat-fixed slide over the boiling water bath on the hot plate **(figure 12.2)**. Alternatively, your instructor may have you set up a ring stand and Bunsen burner to steam your slides.
8. Cover the smear with a small piece of paper towel. The paper towel should just cover the smear and not overlap the edges of the slide.
9. Soak the paper towel/smear with Ziehl-Neelsen carbol fuchsin.
10. Steam the slide for at least 5 minutes. Add more stain as the carbol fuchsin evaporates from the slide. Do not allow the slide to dry out. Heating the slide increases the fluidity of acid-fast cell walls and allows the carbol fuchsin to penetrate the cells.
11. After steaming for 5 minutes, remove the slide. Place the slide on the staining tray and allow the slide to cool.
12. Remove the paper towel and throw it in the trash can.
13. Tip the slide and rinse with distilled water. You may need to wash both sides of the slide to remove excess stain.
14. Decolorize the slide with acid alcohol. Decolorize until the acid alcohol dripping off of the slide is a light red color; this should take about 30 seconds.
15. Counterstain the slide with methylene blue for 3 minutes.
16. Rinse the slide with distilled water and blot dry.
17. View the slide using oil immersion. Acid-fast cells will stain a bright red or fuchsia color. Non-acid-fast cells will appear blue.
18. Discard used dyes according to the chemical safety guidelines at your institution.

## Procedure (Kinyoun)

1. Obtain a clean slide.
2. Prepare your smear as just described in steps 2 to 6 of the Ziehl-Neelsen procedure.

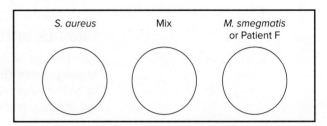

**Figure 12.1 Smear preparation.** Add three drops of water to create smears on your slide.

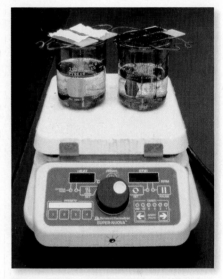

**Figure 12.2 Staining setup for the Ziehl-Neelsen method of acid-fast staining.** Place heat-fixed smears above boiling water and then apply Ziehl-Neelsen carbol fuchsin. Steam the slides for at least 5 minutes.
©Steven D. Obenauf

3. Place the heat-fixed slide on the staining tray and cover the smear with Kinyoun carbol fuchsin. Allow the slide to stain for 3 minutes.

4. Rinse with water; then decolorize the slide with acid alcohol until the acid alcohol running off of the slide is nearly colorless.

5. Again rinse the slide with water.

6. Counterstain with methylene blue for 2 minutes.

7. Rinse with water, blot dry, and view using oil immersion. Acid-fast cells will stain a bright red or fuchsia color. Non-acid-fast cells will appear blue.

8. Discard used dye appropriately.

## Results and Interpretation

The acid-fast stain is used primarily to visualize members of the genus *Mycobacterium* **(figure 12.3).** Members of this genus, particularly *Mycobacterium tuberculosis,* are clinically very important. The results of the acid-fast staining procedure are shown in the figure.

(a)                                                    (b)

**Figure 12.3** **Acid-fast stained mixture of** *Mycobacterium* **and** *S. aureus.* Mycobacteria are slender, acid-fast cells. They appear bright red or fuchsia-colored in an acid-fast stain **(a).** The non-acid-fast cells (*S. aureus*) appear blue **(b).** ©Susan F. Finazzo

Name _____

Date _____

# Acid-Fast Stain

## Your Results and Observations

Record the staining results you observed below. Be sure to include the organism or specimen name and the magnification at which you viewed the slide.

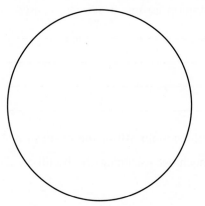 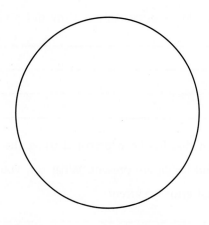 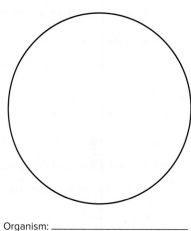

Organism: _____     Organism: _____     Organism: _____

Magnification: _____     Magnification: _____     Magnification: _____

## Interpretation and Questions

1. Why was it important to steam the slides in the Ziehl-Neelsen method?

_____

_____

_____

_____

2. Why is the acid-fast stain clinically important?

_____

_____

_____

_____

3. Which of the stained organisms was acid-fast?

_____

_____

_____

_____

**4.** The patient in the case file had to undergo a 4-month course of treatment. A typical course of treatment for a bacterial infection is 10 days. What does this suggest about mycobacterial diseases? Noncompliance is a major issue when treating TB because of the length of time of the treatment and the population typically exposed to the disease. What potential issues or problems does noncompliance pose in the area of public health?

_____

_____

_____

_____

**5.** The patient initially complained of pain in his hip and knee. A typical diagnosis for someone in this age range and with these symptoms would be osteoarthritis. Why did his physician assume a different cause?

_____

_____

_____

_____

**6.** You are viewing a clinical specimen that has been prepared using the acid-fast stain. All of the organisms are stained blue, and both cocci and bacilli are present. What can you conclude? Assuming the bacilli are mycobacterial cells, what procedural error occurred?

_____

_____

_____

_____

# Endospore Stain
# (Schaeffer-Fulton Method)

## CASE FILE

A 4-month-old boy was brought to the emergency room. His
mother was concerned that the child had not had a bowel
movement in 3 days, had a weak cry, and was eating poorly.
The doctor observed that the child was drooling excessively,
had drooping eyelids, and exhibited general muscle weakness.
Upon further investigation, it was learned that the child was
being fed a commercially available formula and a commercially
available infant cereal sweetened with natural, unpasteurized
honey. The child's symptoms are consistent with infant botulism. Natural
honey has been implicated in food-borne cases of botulism.

©Antonprado/Getty Images RF

A brain scan and nerve conduction test were performed to eliminate
other possible neurological causes. The parent was asked to submit a
sample of the honey and other food products ingested by the child during the previous 2-week period. A sample of honey and a stool sample
from the child were sent to the local public health agency for testing.
Part of that testing of the honey included the spore stain that you will
be performing in this exercise as well as testing the stool for toxin.

Pending a diagnosis, the child was admitted to the intensive care
ward. The patient's breathing was monitored closely. He was given
intravenous fluids, and antitoxin (botulism immune globulin intravenous, or
BIG-IV) was administered. The patient did show improvement and
remained hospitalized for 2 weeks.

## LEARNING OUTCOMES

At the completion of this exercise, students should be able to
- list the common genera of endospore-forming bacteria.
- explain the significance of endospore formation to bacteria.
- correctly perform an endospore stain.
- explain the basis for the steps in the staining technique.
- postulate on the possible impact of endospore-forming bacteria
  on human health.

## Background

Bacteria are subject to their environment. They must compensate for
changing environmental conditions or die. Two special genera have found

a way to protect themselves from adverse environmental conditions. When exposed to conditions such as nutrient limitation or waste accumulation, **vegetative cells** (actively metabolizing) of members of the genera *Bacillus* (aerobe) and *Clostridium* (anaerobe) begin the process of **sporulation** and eventually form **endospores.** An endospore is a survival structure formed within the bacterial cell wall that allows the cell to survive until environmental conditions improve. The endospore contains a copy of the organism's chromosome, a little cytoplasm, and a small amount of cell machinery. The spore coat is composed of keratin and other proteins, calcium, dipicolinic acid, and peptidoglycan.

Vegetative cells–when stressed by such factors as nutrient limitation, temperature changes, and waste accumulation–begin the process of sporulation **(figure 13.1).** Sporulation starts with the replication of the bacterial chromosome and the spatial separation of the chromosomes in different regions within the cell's cytoplasm. Separation of the chromosomes is accomplished by the formation of a wall, or septum, within the cell. The region on one side of the septum is called the forespore and will eventually mature to become the endospore. The region on the other side of the septum is called the sporangium. The sporangium is metabolically active and produces and deposits the compounds necessary for the completion of the spore coat layers. The sporangium dies once the endospore matures. The mature endospore, or free spore, is highly environmentally resistant and can exist as a resting spore for decades. However, once environmental conditions are conducive for growth and reproduction, the endospore can germinate and release a metabolically active vegetative cell.

The location of the endospore within the cell varies with the bacterial species. An endospore that develops at the end of a cell is called a **terminal** endospore. Those that form in the middle of the cell are called **central** endospores, and endospores that are located between the center and the end of the cell are called **subterminal** endospores. The endospore matures, and over time the cell dies. The cell wall breaks down, releasing the endospore. It is now just called a spore, or free spore.

Endospores are environmentally resistant and very difficult to destroy. They resist destruction by heat, radiation, desiccation, and chemicals. These properties, which preserve the organism, also make the endospore very difficult to stain. These characteristics also explain the potential use of endospores as biological weapons. Anthrax is caused by *Bacillus anthracis.* Inhaled aerosolized anthrax spores germinate very quickly to form vegetative cells in the lungs and can cause anthrax. Because anthrax spores, when dispersed in soils and on surfaces, can remain infective for many years, they represent a serious environmental and health concern. Scientists have postulated that endospores can remain viable for thousands of years.

The endospore-forming genus *Clostridium* also includes several disease-causing species. Members of this genus include the causative agents of tetanus (*Clostridium tetani*), gas gangrene (*Clostridium perfringens*), and botulism (*Clostridium botulinum*). The toxins produced by these organisms lead to a myriad of symptoms and signs, including spastic paralysis (tetanus); myonecrosis, or necrotizing myositis (gas gangrene); and flaccid paralysis (botulism).

The endospore stain is classified as a differential (or structural) stain. Heat-fixed smears are steamed over boiling water with the primary stain, malachite green. Heating the endospores with the primary stain allows the stain to penetrate and dye the spore coat. The slide is then cooled. As the spore cools, the stain is trapped inside. Smears are decolorized and then counterstained with a secondary stain (safranin) to dye the vegetative cells.

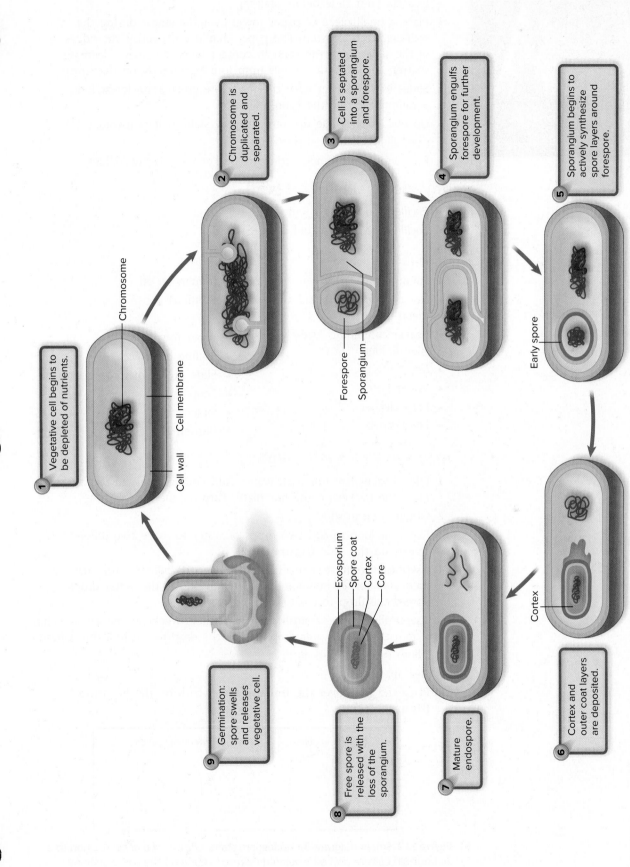

**Figure 13.1** The sporulation cycle in bacteria follows the conversion of an actively metabolizing vegetative cell to the resting, or dormant, form of a bacterial endospore.

1. Vegetative cell begins to be depleted of nutrients.

Chromosome

Cell wall    Cell membrane

2. Chromosome is duplicated and separated.

3. Cell is septated into a sporangium and forespore.

Forespore
Sporangium

4. Sporangium engulfs forespore for further development.

5. Sporangium begins to actively synthesize spore layers around forespore.

Early spore

6. Cortex and outer coat layers are deposited.

Cortex

7. Mature endospore.

Exosporium
Spore coat
Cortex
Core

8. Free spore is released with the loss of the sporangium.

9. Germination: spore swells and releases vegetative cell.

©Steven D. Obenauf

## Tips for Success

- Heat-fix your slide before staining.
- Place a small piece of paper towel over the smear during the staining/steaming step. The paper should not overlap the edges of the slide. Keep the stain in contact with the smear, decrease spatter, and prevent crystals from forming directly on the smear.
- Slides **MUST** steam with the malachite green stain for at least 5 minutes to dye the spores.
- Do not let the slide dry out. Reapply stain as it evaporates.
- Cool slides before decolorizing.
- Remove steamed slides from the beaker with a slide holder.

## Organisms (Plate or Slant)

- *Bacillus* spp., 24-hour culture
- *Bacillus* spp., 1-week-old culture

## Materials

- Staining tray
- Hot plate or ring stand and metal mesh
- Beaker containing water with slide holder
- Malachite green
- Water bottle
- Lens cleaner
- Lens paper

- Immersion oil
- Bibulous paper
- Safranin
- Paper towel
- Boiling beads
- Slide holder or forceps
- Loop
- Slide
- Bunsen burner

## Procedure (Schaeffer-Fulton)

1. Fill a beaker half full with water. Add boiling beads to the beaker. Place the beaker on the hot plate. Turn on the hot plate.
2. Obtain a clean slide.
3. Use your loop to add two drops of water to make two different smears on the slide **(figure 13.2)**.
4. Aseptically remove a small amount of the *Bacillus* culture that has incubated for 24 hours and mix it into one of the water drops. Spread the drop to create a thin smear.
5. Aseptically remove a small amount of the 1-week-old culture and mix it into the other water drop. Spread the drop to create a thin smear.
6. Allow the smears to air-dry.
7. Heat-fix the smear.
8. Place the slide over the **boiling** water bath on the hot plate **(figure 13.3)**.

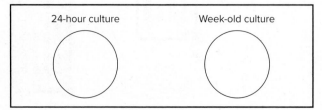

**Figure 13.2 Smear diagram for endospore stain.** Comparison of sporulation in a 24 hour-old culture and a 1-week-old culture is done on the same slide by creating two smears and then heat-fixing and staining the slide.

9. Cover the smear with a small piece of paper towel. The paper towel should just cover the smear and not overlap the edges of the slide.

10. Soak the paper towel/smear with malachite green.

11. Steam the slide for at least 5 minutes. Add more stain as the malachite green evaporates from the slide. Do not allow the slide to dry out. Heating the slide allows the malachite green to penetrate the spore.

12. After steaming for 5 minutes, remove the slide using the slide holder. Place the slide on the staining tray and allow the slide to cool.

13. Use forceps to remove the paper towel and throw it in the trash can.

14. Tip the slide and rinse with distilled water. You may need to wash both sides of the slide to remove excess stain.

15. Counterstain the slide with safranin for 1 minute.

16. Rinse the slide with distilled water and blot dry.

17. View the slide using oil immersion. Endospores/spores will stain green to blue green. Vegetative cells will appear red.

## Results and Interpretation

Bacterial endospores are produced when cells are stressed. The endospore is environmentally resistant, clinically important, and difficult to visualize with normal staining techniques, like the Gram stain. Heating or steaming the smear with the primary stain, malachite green, enhances the penetration of the stain into the endospore. Malachite green imparts the blue-green color **(figure 13.4)** observed in spores stained with this technique. Safranin is used as the counterstain. Vegetative cells will appear red **(figure 13.5)**.

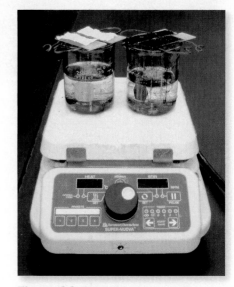

**Figure 13.3 Steaming water bath.** Slides must be heated for at least 5 minutes for the staining method to be effective. A steaming water bath provides an effective means to heat the slide. ©Steven D. Obenauf

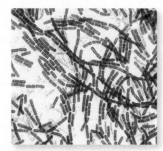

**Figure 13.5** Vegetative cells of *Bacillus.* ©Steven D. Obenauf

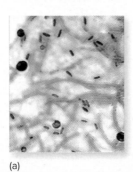 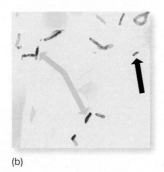

(a)          (b)

**Figure 13.4 Vegetative cells and endospores.** Both of these images are of endospore stained slides. The light blue arrows are pointing to vegetative cells. The blue-green structures are bacterial spores. The yellow arrows are pointing to free spores, and the black arrows are pointing to endospores. ©Steven D. Obenauf

# NOTES

Name _____

Date _____

# Endospore Stain

## Your Results and Observations

Record your staining results below. Indicate the organism's name, the culture age, and the magnification at which you observed your stained slides.

24-hour culture

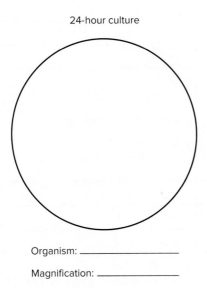

Organism: _____

Magnification: _____

1-week-old culture

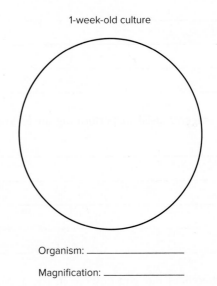

Organism: _____

Magnification: _____

## Interpretation and Questions

**1.** You forgot to heat-fix your slide. What would you expect to see?

_____

_____

_____

_____

_____

**2.** Why might pasteurizing honey not be enough to make it completely safe for infants?

_____

_____

_____

_____

**3.** Only two genera produce endospores; name those genera and give one reason each genus is either medically or environmentally important.

_____

_____

_____

_____

_____

**4.** How would your slide (cells and endospores) have appeared if you had performed a Gram stain rather than an endospore stain?

_____

_____

_____

_____

_____

**5.** List the reagents used in performing an endospore stain and describe their function.

_____

_____

_____

_____

_____

**6.** What differences did you observe in the appearance of the endospore stained slides of the 24-hour-old culture and 1-week-old culture?

_____

_____

_____

_____

_____

# Blood Agar

©Dr P. Marazzi/Science Source

## LEARNING OUTCOMES

At the completion of this exercise, students should be able to

- understand the function of blood agar.
- describe the types of hemolysins and their function.
- interpret the type of hemolysis associated with growth on a blood agar plate.

## Background

Blood agar is a medium that is both **enriched** and **differential.** Many bacteria can grow on this medium. The iron and other extra nutrients blood provides are needed by fastidious streptococci and a number of other pathogenic bacteria in order to grow. Some bacteria simply grow on it, whereas others partially or completely break down the blood cells embedded in the agar. The ability to break down blood cells (or not) makes this medium differential. Diagnostically, plates containing 5% to 10% sheep blood are used most often.

    *Streptococcus* and *Staphylococcus* both make toxins called **hemolysins** that can break down red blood cells (RBCs). Alpha-hemolysins cause partial lysis of RBCs, which lead to darkening of the blood agar around the colony called **alpha-hemolysis.** Hemoglobin breakdown causes the green color often seen in alpha-hemolysis. Beta-hemolysins cause complete lysis of the RBCs and breakdown of hemoglobin, which leads to colorless clearing of the blood agar around the colony called **beta-hemolysis.** If an organism causes no hemolysis, it is reported as **gamma-hemolysis.**

There are many different species of *Streptococcus*, including *Streptococcus pneumoniae* (the cause of pneumococcal pneumonia and a potential cause of meningitis or otitis media) and *Streptococcus mutans* (a major cause of tooth decay). *Enterococcus faecalis* (formerly known as *Streptococcus faecalis*) is one of the leading causes of hospital-acquired infections. Another streptococcal organism that many people have unhappily encountered during their lives is *Streptococcus pyogenes*, which can cause **bacterial pharyngitis,** or "strep throat." The ability of *S. pyogenes* to cause disease is due in large part to its ability to resist phagocytosis and to make toxins and enzymes, including types of hemolysins called streptolysins. *S. pyogenes* will cause beta-hemolysis on a blood plate (and, in fact, is often called beta-hemolytic strep); another major streptococcal pathogen, *S. pneumoniae*, will cause alpha-hemolysis. *E. faecalis* causes gamma-hemolysis. It is important to treat streptococcal pharyngitis (usually with penicillin) to avoid some potentially dangerous complications, including rheumatic fever, scarlet fever, and glomerulonephritis. On the other hand, viruses are the most common cause of pharyngitis. Treating viral infections with antibacterial drugs isn't a good idea. Although enzyme immunoassays or other immunoassays are typically used for the diagnosis of strep throat, blood plates are still considered the "gold standard" for diagnosing bacterial pharyngitis because they typically have a lower rate of false negatives than immunoassays.

###  Tips for Success

- The key to reading results on a blood agar plate is not whether the organism grows—they all will. Look for changes in the blood agar on the **edge** of the colony. Is there a slight change, a big change, or no change?

## Organisms (in Broth Cultures—Handle These Pathogens with Caution)

- *Streptococcus pyogenes*
- *Streptococcus sanguinis (sanguis)*
- *Streptococcus (Enterococcus) faecalis*
- Patient A sample for diagnosis

## Materials (per Team of Two or Four)

- 1 blood agar plate
- Inoculating loop
- Parafilm or tape (optional)
- Bunsen burner or microincinerator

## Procedure

### Period 1

1. Use your marker to divide the plate into quarters (mark the bottom of the plate as shown in **figure 14.1**) and label each quarter with the name of the organism or sample that will be inoculated there.
2. Streak each organism as shown in the diagram. Your professor may also have you stab the agar in several areas along the streak inoculation with the loop or needle (this will enhance hemolysis with organisms that produce hemolysins that are active only in low oxygen conditions).

©Steven D. Obenauf

**Figure 14.1** Inoculating the plate.

3. Put the plate lid side down in the incubator. You may use tape or Parafilm to keep the edges shut. Incubation should be for 24 to 48 hours at 35°C to 37°C. Excessive incubation can cause the blood to turn brown.

4. Handle and dispose of your *Streptococcus* and Patient A cultures appropriately.

## Period 2

1. Observe and record the growth on your plate. Blood agar plates are best read by looking through them with light behind the plate to see the changes in the blood around the growth **(figure 14.2)**. Look around the **edge** of the colonies on the plate **(figure 14.3)**.

2. If the alpha-hemolysis or beta-hemolysis has not fully developed, you may need to put the plate back in the incubator and read it again at next lab period.

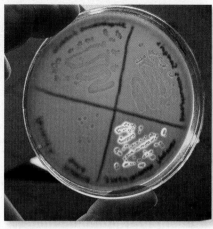

**Figure 14.2** Reading the plate.
©Steven D. Obenauf

## Results and Interpretation

**Figure 14.4** will show you what to look for and what each type of hemolysis should look like. Compare your Patient A sample to the three *Streptococcus* samples on the plate. **Alpha-hemolysis (figure 14.4a)** will appear as a "halo" or partial clearing around the bacteria. It is usually very narrow and sometimes greenish. **Beta-hemolysis (figure 14.4b)** will appear as a clear zone around the bacteria. The blood is completely digested–all that is left is agar. It often starts narrow and becomes wider. **Gamma-hemolysis (figure 14.4c)** will show as no change at all in the blood around the bacteria. The bacteria are growing but not making anything to damage the blood.

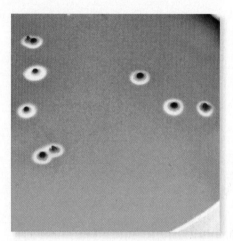

**Figure 14.3** Colonies and hemolysis.
©Steven D. Obenauf

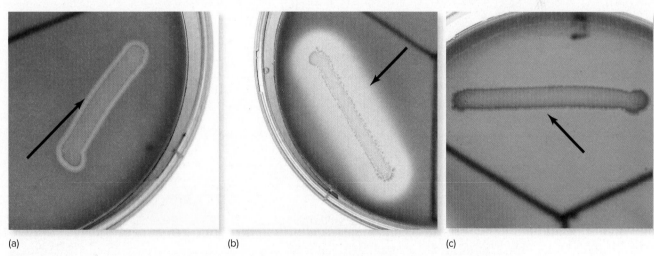

(a)                    (b)                    (c)

**Figure 14.4** **(a)** Alpha-hemolysis; **(b)** beta-hemolysis; **(c)** gamma-hemolysis. ©Steven D. Obenauf

# NOTES

# Blood Agar

## Your Results and Observations

Draw your results or photograph with your mobile device and attach or email a photograph.

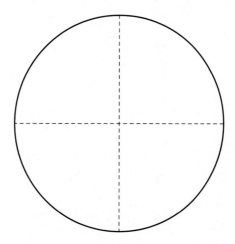

## Interpretation and Questions

| Organism or sample | Result (appearance of the agar around the growth/ breakdown of the blood) | Hemolysis type (alpha, beta, or gamma) |
|---|---|---|
| *Streptococcus pyogenes* | | |
| *Streptococcus sanguinis* | | |
| *Enterococcus faecalis* | | |
| Patient A sample | | |

**1.** Do all bacteria that grow on blood agar break down the blood? Explain why just about all bacteria will grow on blood agar. How is it a differential medium?

_____

_____

_____

_____

**2.** What type of hemolysis does the bacteria infecting Patient A cause? Based on that and the information in the Background section, with which bacteria are they probably infected?

_____

_____

_____

_____

# Mannitol Salt Agar

## CASE FILE

In early summer, a festival was held in a community 50 miles south of Dallas, Texas. Within hours of attending, 98 people became ill with severe nausea, vomiting, and diarrhea. Seventeen of them sought medical attention. Most had recovered by the next day. Testing of various samples of food by the health department found evidence that toxin-producing *Staphylococcus aureus* may be present in ham from one of the food vendors at the event. You'll be testing Sample M to help find out if *S. aureus* caused the symptoms in the ill people.

Source: Janice Haney Carr/Centers for Disease Control and Prevention

### LEARNING OUTCOMES

At the completion of this exercise, students should be able to

- understand the selective and differential functions of mannitol salt agar.
- interpret growth on a mannitol salt agar plate and determine the bacterium it represents.
- describe the use of mannitol salt agar in the identification of *Staphylococcus aureus*.

## Background

Of all the disease-causing bacteria, the *Staphylococci* are some of the most talented. *Staphylococcus aureus* is the most pathogenic of the staphylococci. It can infect the skin and many other tissues in the body. It can produce a wide range of harmful enzymes and toxins. Among these are enzymes and toxins that kill white blood cells, trigger a fibrin clot, or cause shock. As antibiotic-resistant strains of staphylococci, such as methicillin-resistant *S. aureus* (MRSA), and vancomycin-resistant *S. aureus* (VRSA), continue to spread, the infections that they cause are becoming increasingly difficult to treat. Even worse, *S. aureus* doesn't even have to be inside of you to make you sick. It can release an enterotoxin as it grows in food that will make an individual sick within hours of eating.

Mannitol salt agar is an example of a medium that is both **selective (figure 15.1)** and differential. This medium contains a **high concentration of salt** (7.5% NaCl), which makes the medium selective by inhibiting the growth of most bacteria but allowing salt-tolerant (halophilic)

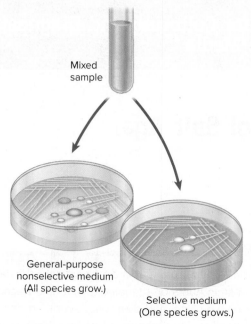

Mixed
sample

General-purpose
nonselective medium
(All species grow.)

Selective medium
(One species grows.)

**Figure 15.1** Function of a selective medium.

staphylococci to grow. In addition, this medium contains the sugar **mannitol** and **pH indicator** phenol red. Disease-causing staphylococci ferment the sugar mannitol. Mannitol fermenters release acid into the medium, which turns the phenol red in the medium around the colony yellow as the pH changes. Non-disease-causing staphylococci (like *Staphylococcus epidermidis*) grow but do not change the medium's color. The combination of mannitol and phenol red allows for differentiation between pathogenic and nonpathogenic staphylococci.

## ✓ Tips for Success

- Make **short** streaks, so that the color zones from adjacent samples don't merge and make it hard to read your results.
- Incubating plates for more than 24 hours may cause zones around the colonies to become too large and interfere with the results. Removing and/or refrigerating plates may be needed.

## Organisms (in Broths)

- *Staphylococcus aureus*
- *Staphylococcus epidermidis*
- *Escherichia coli*
- Sample M

## Materials (per Team of Two or Four)

- 1 mannitol salt agar plate
- Inoculating loop
- Bunsen burner or microincinerator
- (Optional) 1 tryptic soy agar (TSA) plate

**Figure 15.2** Marking and inoculating the plate. ©Steven D. Obenauf

## Procedure

### Period 1

1. Use your marker to divide the plate into quarters (mark the bottom of the plate as shown in **figure 15.2**); label each quarter with the name of the organism that will be inoculated there. As an optional activity, the organisms can also be inoculated onto a control TSA plate, which is not selective.

2. Make a short streak of each organism.

3. Put the plate lid side down in the incubator.

### Period 2

1. Observe the growth on your plate and record the results in your lab report.

## Results and Interpretation

Mannitol salt plates can be read against either a dark or light background. First, determine if the organism in each area grew. Second, did the color of the medium around the growth change? **Figure 15.3a** to **c** shows possible results. An organism can grow and produce acid, grow and not produce acid, or not grow at all.

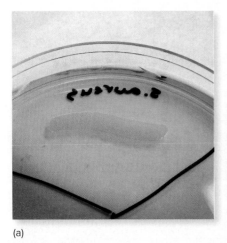

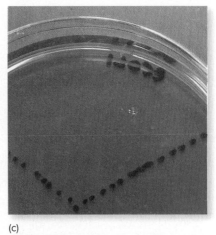

(a)                                (b)                                (c)

**Figure 15.3 Growth on mannitol salt agar. (a)** Pathogenic *Staphylococcus aureus* growing on mannitol salt agar, fermenting mannitol, producing acid, and changing the color of the pH indicator. **(b)** *Staphylococcus epidermidis* growing but not fermenting mannitol and not producing acid. **(c) Lack of growth** due to inhibition by the high level of salt in medium. ©Steven D. Obenauf

# NOTES

108

Name _____

Date _____

# Mannitol Salt Agar

## Your Results and Observations

Draw your results or photograph them with a digital camera or mobile device.

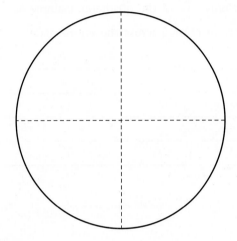

## Interpretation and Questions

| Organism | Growth (1/2) | Appearance of the growth or the agar around the growth |
|---|---|---|
| *Staphylococcus aureus* | | |
| *Staphylococcus epidermidis* | | |
| *Escherichia coli* | | |
| Sample M | | |

**1.** Based on the results seen on your plate, does Sample M contain *Staphylococcus aureus*? Why or why not?

_____

_____

_____

_____

_____

**2.** Take a look at the Background section. Would this medium still work (be selective for *S. aureus*) if it did not have salt in it? Why or why not?

_____

_____

_____

_____

_____

**3.** Your partner makes a very, very long streak in each quadrant. The inverted plates are placed in the incubator on Thursday. In class the next Tuesday, you remove your plate and the entire plate is yellow, even though there is growth in only two of the quadrants (Sample M, *S. aureus*) on opposite sides of the plate. How do you explain the color change across the entire plate?

_____

_____

_____

_____

_____

# Eosin Methylene Blue Agar

## CASE FILE

Two water samples have arrived in a testing lab. One is from a group of homeowners with homes on the same street, sharing a common well. A new resident of the street is concerned about water quality. The second is from individuals concerned about the safety of the water in a river near their homes.

Drinking water is not sterile. It contains bacteria. The key thing to know is what type. Normally, the bacteria found in water supplies are harmless. Finding a type of bacteria known as coliform bacteria, and in particular, *Escherichia coli*, can be a cause for concern. Livestock, inadequate wastewater treatment plants, leaky septic systems, landfills, and stormwater runoff are common sources of fecal contamination of water supplies, including rivers, lakes, and wells. While water used for swimming or boating ("recreational water") is allowed to have relatively small amounts of *E. coli* (the EPA standard for primary contact water is 235 *E. coli* per 100 milliliters in a single sample) in addition to the other bacteria found in that water, the standard for drinking water is zero. So if any *E. coli* are detected in drinking water, it fails the standard. The phrase "one is too many" applies here. The eosin methylene blue (EMB) agar that you will use in this exercise is used as part of the process of confirmatory testing for *E. coli* in water.

Source: CDC/Peggy S. Hayes & Elizabeth H. White, M.S.

### LEARNING OUTCOMES

At the completion of this exercise, students should be able to
- understand the selective and differential functions of EMB agar.
- examine the various colors of growth on an EMB agar plate and determine the category of bacteria they represent.

## Background

Water can be a dangerous thing. A wide variety of serious microbial diseases can be transmitted by contact with or ingestion of water. These include cholera, shigellosis, amoebiasis, giardiasis, schistosomiasis, hepatitis A, and viral gastroenteritis.

Coliform bacteria can live in the environment or the intestines of many animals (including us). Fecal coliform bacteria such as *Escherichia coli* live primarily or exclusively within the intestines of warm-blooded and cold-blooded animals. Therefore, the presence of these bacteria in water or food

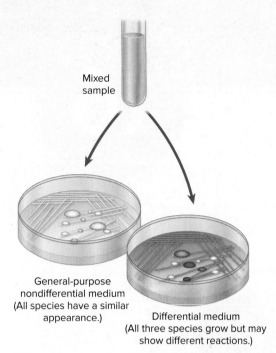

Mixed sample

General-purpose nondifferential medium (All species have a similar appearance.)

Differential medium (All three species grow but may show different reactions.)

**Figure 16.1** Function of a differential medium.

(a)

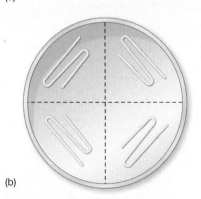

(b)

**Figure 16.2 (a)** Uninoculated plate and **(b)** pattern for marking and inoculating the plate.
©Steven D. Obenauf

may indicate possible fecal contamination. When testing water safety, we usually do not look directly for all the possible water-borne pathogens such as *Giardia* or hepatitis A virus. Instead, we test for the presence and quantity of fecal coliforms. They are used as **indicators** of the presence of other harmful microorganisms that pose significant threat to humans. In exercise 19, "Plate Count," we will perform a commonly used test for quantitating bacteria, but it provides little information that can be used to identify them. The higher the level of fecal coliforms present in a water sample, the greater the risk that something harmful will also be there. *E. coli* is a particularly good indicator because it is *not* typically found living in soil or water, so its presence is more strongly correlated with fecal contamination than other bacteria.

A standard qualitative test for determining water safety is the fecal coliforms test. Since coliform bacteria ferment the sugar lactose, the first **(presumptive)** step in the fecal coliforms test looks for lactose-fermenting organisms in the sample. If that step is positive, a second **(confirmatory)** test looks more specifically for fecal coliform bacteria. One of these confirmatory tests is the use of eosin methylene blue agar. A number of other media in addition to EMB can be used to test for the presence of *E. coli*.

Eosin methylene blue agar is an example of a medium that is both selective and **differential (figure 16.1).** The two dyes together–eosin and methylene blue–make the medium selective by inhibiting the growth of gram-positive bacteria but allowing many gram-negatives (including coliforms) to grow. In addition, this medium contains the sugar lactose, which allows for **differentiation** between non-lactose fermenters, lactose fermenters, and heavy lactose fermenters. It is especially useful for identifying fecal coliform bacteria. The production of acid while fermenting lactose will trigger a pink to metallic green color change that will distinguish coliform bacteria from other gram-negatives and *E. coli* from other coliforms.

## ✔ Tips for Success

- The way to read EMB agar plates is different from blood, MacConkey, or mannitol salt agar plates. EMB agar plates are best read with a dark background underneath them–do not hold them up and look through them or you will not see the colors of the colonies well.

## Organisms (in Broths)

- Sample P (water) or student water sample
- *Escherichia coli*
- *Enterobacter aerogenes*
- *Pseudomonas aeruginosa*

## Materials (per Team of Two or Four)

- 1 EMB agar plate
- Inoculating loop
- Bunsen burner and striker

## Procedure

1. Interested students should get a sample bottle or container to obtain water from a river, canal, lake, or other freshwater source.

## Period 1

1. Use your marker to divide the plate into quarters (mark the bottom of the plate as shown in **figure 16.2**); label each

quarter with the name of the organism or sample that will be inoculated there.

2. Streak each organism as shown in the diagram or with a single straight streak.

3. Put the plate lid side down in the incubator.

## Period 2

1. Observe and record the growth on your plate. Look at the color of the growth or colonies on the plate, not the medium **(figure 16.3)**.

## Results and Interpretation

Read your plates against a dark background (for example, your lab bench-top)–you may need to lift the lid a bit. Gram-positive bacteria will grow poorly on this medium or not at all **(figure 16.4a)**. Gram-negative bacteria that are not coliforms will grow well but are basically colorless **(figure 16.4b)**. Coliform bacteria (lactose fermenters) will grow well and the colonies will be pink with purple in the center **(figure 16.4c)**. Rapid or strong lactose fermenters will grow well and will appear dark purple with a distinct metallic green color **(figure 16.4d)**.

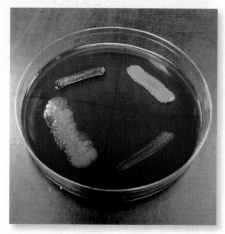

**Figure 16.3** Growth on the plate.
©Steven D. Obenauf

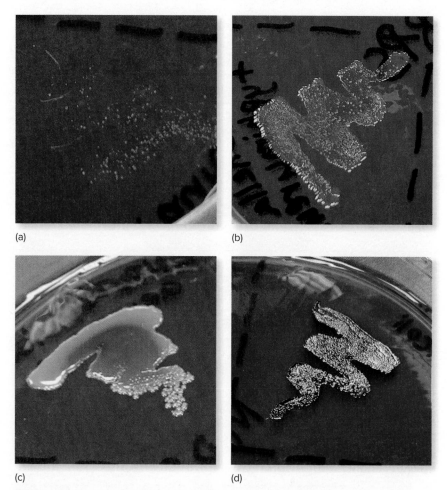

(a)

(b)

(c)

(d)

**Figure 16.4 Growth on EMB agar. (a) Gram-positive organism** growing poorly due to inhibition by methylene blue in medium. **(b) Gram-negative** non-lactose fermenter. **(c) Coliform** lactose fermenter with mucoid pink and purple growth. **(d)** *E. coli* with green sheen caused by rapid lactose fermentation. ©Steven D. Obenauf

# NOTES

114

Name _____

Date _____

# Eosin Methylene Blue Agar

## Your Results and Observations

Draw your results or photograph with your mobile device and attach or email a photograph.

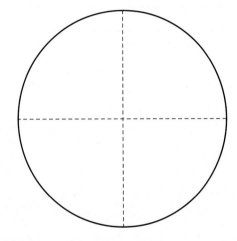

| Organism or sample | Result (appearance of the growth) | Lactose fermenter/ heavy fermenter | Non-lactose fermenter |
|---|---|---|---|
| *Escherichia coli* | | | |
| *Enterobacter* | | | |
| *Pseudomonas* | | | |
| Sample P (lake water) | | | |

## Interpretation and Questions

1. What bacteria or types of bacteria are in Sample P or the student sample?

_____

_____

_____

_____

**2.** You are a public health official analyzing these results. Do you think that harmful organisms such as *Giardia* or hepatitis A virus could be in the water? Why or why not?

_____

_____

_____

_____

**3.** Name the differential agent included in EMB and describe how it differentiates between groups of bacteria.

_____

_____

_____

_____

# MacConkey Agar

**LEARNING OUTCOMES**

At the completion of this exercise, students should be able to

- explain how MacConkey agar is both selective and differential.
- interpret the various colors of growth on a MacConkey agar plate and determine the category of bacteria they represent.
- explain when and how MacConkey agar would be used in a clinical setting.

## Background

A hospital can be a dangerous place to stay. Nosocomial, or **hospital-acquired, infections** present a significant risk to patients and cause well over a million illnesses and around 100,000 deaths each year. Urinary tract infections, blood infections, and pneumonias are the three most commonly seen categories of nosocomial infections. A type of hospital-acquired pneumonia seen in patients on mechanical ventilators is known as ventilator-associated pneumonia (VAP). The most commonly seen causes of VAP include the gram-positive cocci *Streptococcus* and *Staphylococcus* (including MRSA). Gram-negative causes include several coliform bacteria: *Klebsiella, Enterobacter,* and *Escherichia coli* as well as non-coliform bacteria such as *Pseudomonas, Acinetobacter,* and *Serratia.* The fungus *Candida* is also a possibility.

MacConkey agar is an example of a medium that is both selective and differential. Bile salts and the dye crystal violet make the medium **selective**

**Figure 17.1 Marking and inoculating the plate.** Divide the plate into quadrants and then streak each quadrant with one of the organisms just listed.

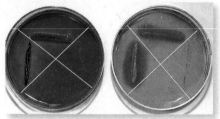

**Figure 17.2 Growth on EMB and MacConkey agars.** EMB agar is shown on the left and MacConkey agar is shown on the right. The top quadrant on each plate was inoculated with ICU patient sample, the left quadrant on each plate was inoculated with *E. coli,* the bottom quadrant of each plate was inoculated with *Pseudomonas aeruginosa,* and the right quadrant with *Staphylococcus aureus.* **Gram-positive** organisms grow poorly on these media due to inhibition by bile salts and crystal violet. **Gram-negative** non-lactose-fermenting bacteria grow as colorless colonies. **Coliform** bacteria form pink colonies. ©Susan F. Finazzo

**Figure 17.3 A comparison of the growth of *E. coli* on MacConkey and EMB agars.** *E. coli* is a gram-negative, lactose-fermenting coliform. When growing on MacConkey agar (left), *E. coli* produces a pink colony and causes pink dye to be deposited in the agar surrounding the colony. When growing on EMB agar (right), colonies of *E. coli* exhibit a metallic green color. ©Susan F. Finazzo

by inhibiting the growth of gram-positive bacteria but allowing many gram-negatives (including coliforms) to grow. In addition, this medium contains the sugar lactose and the pH indicator neutral red, which allows for **differentiation** between non-lactose fermenters such as *Pseudomonas* and lactose fermenters such as *Klebsiella* and *Escherichia.* It is especially useful for identifying enteric bacteria. Acid is produced and released into the medium during the fermentation of lactose. In response to the lower pH, neutral red turns from pink to red. The color change distinguishes lactose-fermenting coliform bacteria from other gram-negative bacteria. Colonies of lactose fermenters will appear pink. Non-lactose-fermenting gram-negative organisms will utilize peptone in the medium for growth. Their colonies will be clear or uncolored. There are versions of MacConkey agar with different chemical makeups that are used to identify *Enterococcus* or inhibit *Proteus.*

## ✅ Tips for Success

- If you are doing this exercise at the same time as other similar ones, such as mannitol salts or EMB agar, make sure to label your plates as you pick them up. They are similar in color.

## Organisms (in Broths)

- Sample I (ICU patient sample)
- *Escherichia coli*
- *Staphylococcus epidermidis*
- *Pseudomonas aeruginosa*

## Materials (per Team of Two or Four)

- 1 MacConkey agar plate
- 1 nutrient agar plate (optional)
- Inoculating loop
- Bunsen burner and striker

## Procedure

### Period 1

1. Use your marker to divide the plate into fourths (mark the bottom of the plate as shown in **figure 17.1**); label each quarter with the name of the organism or sample that will be inoculated there.
2. Streak each organism as shown in the diagram or with a single straight streak.
3. Put the plate lid side down in the incubator.

### Period 2

1. Observe and record the growth on your plate.
2. Look at the color of the growth or colonies on the plate, and the **medium around them (figures 17.2 and 17.3).**
3. Gram-positive bacteria will grow poorly on this medium or not at all (figure 17.2, right quadrant).
4. Gram-negative bacteria that are not coliforms will grow well but are basically colorless (figure 17.2, bottom quadrant).
5. Coliform bacteria (lactose fermenters such as *E. coli* and *Enterobacter*) will grow well and the colonies will be pink or red (figure 17.2, top and left quadrants).
6. Record your results.

## Results and Interpretation

MacConkey and EMB are both selective and differential agars used for the isolation and identification of gram-negative enteric organisms. Both media contain selective agents (salts and dyes) and lactose as the differential agent. See figure 17.3 for a comparison of the two.

Name _____

Date _____

# MacConkey Agar

## Your Results and Observations

Draw your results or photograph them with a digital camera or mobile device.

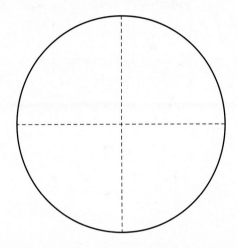

| Organism or sample | Result (appearance of the growth) | Interpretation (what the results indicate) |
|---|---|---|
| *Escherichia coli* | | |
| *Staphylococcus* | | |
| *Pseudomonas* | | |
| Sample I (ICU patient) | | |

## Interpretation and Questions

1. What bacteria or types of bacteria are in Sample I?

_____

_____

_____

_____

**2.** If the staining of the respiratory secretions in the case file had been negative for bacteria and positive for fungi, what would be the likely organism?

_____

_____

_____

_____

**3.** Complete the table for the media studied in this unit. In the last column, indicate if the medium is used to identify a specific organism or genus.

| Medium | Selective (S), differential (D), enriched (E) | Selective agent | Differential agent | Enriched agent | Positive reaction/ reaction of interest | Used to identify |
|---|---|---|---|---|---|---|
| Blood agar | | | | | | |
| Monnitol salt | | | | | | |
| EMB | | | | | | |
| MacConkey | | | | | | |

EXERCISE

## 18

# Chemical Content of Media

### LEARNING OUTCOMES

At the completion of this exercise, students should be able to

- determine which type of medium (defined or undefined) best supports the growth of bacteria.
- compare the growth of fastidious and nonfastidious organisms.
- define and exemplify *fastidious organisms, complex medium,* and *defined medium.*
- explain the use and function of a spectrophotometer.

## Background

Bacteria vary greatly in their nutrient and environmental requirements. The diversity in bacterial growth requirements is reflected in the diversity of media formulations used to support their growth. All bacteria require the element carbon to build organic molecules. An **autotroph** ("feed-self") is an organism that utilizes inorganic carbon ($CO_2$) as its source of carbon. Cyanobacteria are examples of autotrophic members of the domain Bacteria. They use $CO_2$ in the process of photosynthesis to synthesize their own carbohydrates. **Heterotrophic** organisms cannot synthesize organic molecules from $CO_2$. They must use preformed organic molecules in their environment as carbon sources. Most members of the domain Bacteria are considered heterotrophs. The addition of some form of combined carbon to a growth medium is essential for most organisms. Within the heterotrophs, some organisms require more assistance from the environment than others to support their growth. Organisms that require unusual or complex nutrient mixtures to grow are called **fastidious.** Disease-causing bacteria are typically fastidious, as they are adapted to living in the tissues of the human body, using them as a source of nutrients. *Neisseria,* the cause of meningitis (exercise 10) and gonorrhea, is an example of a fastidious genus. These organisms require a nutrient-rich medium high in amino acids and iron. *Neisseria* is typically grown on agar containing heated blood.

In order to cultivate bacteria in the laboratory, we use special media to create the proper environment. The chemical content of different media can vary greatly from one another but are generally placed into one of two categories. Media are classified as either a complex (undefined) medium or a chemically defined medium. A **complex medium** uses preparations of **extracts** from other organisms. Extracts are not chemically definable (that is, the exact amount of each component in an extract is not known). These extracts are often obtained from animals, plants, or yeast. Yeast extract agar and brain-heart infusion agar are both examples of commonly used complex media. Other ingredients like milk and pep-tone (partially degraded protein rich in amino acids) can also be used. Complex or undefined media are rich in nutrients and support a wide range of bacterial growth. A **chemically defined medium** contains pure organic or inorganic compounds added in exact, finite amounts. The exact chemical composition is known for a defined medium, in contrast to com-plex media. A defined medium is only used for a narrow range of bacteria that are less fastidious. More fastidious organisms have very special nutri-ent requirements that may be met by using a complex medium with added nutrients. For example, in order to grow *Streptococcus* in the labora-tory, we often use an **enriched medium** (a medium to which specific nutrients are added to support the growth of fastidious organisms) like blood agar. Blood agar contains blood, which provides necessary nutrients like iron and protein to support the growth of very fastidious *Streptococci.*

How a medium is classified depends on whether we are considering its chemical composition or its use. The media used in some earlier exer-cises (EMB, mannitol salts, and MacConkey) were all complex, undefined media. In addition, they were also classified as selective, differential, or both selective and differential. *Complex* describes the medium's chemical composition. *Selective* and *differential* describe the medium's function. Selec-tive media permit the growth of some organisms and inhibit the growth of others. Differential media distinguish bacteria based on some metabolic reaction. In this exercise, the chemical composition and its relationship to growth are examined.

## Measuring Bacterial Growth

Bacterial growth can be measured either qualitatively or quantitatively. There are instances when qualitative data are adequate. In the differential media exercises, growth (qualitative measurement) along the streak line provided all of the information needed. For example, growth on EMB agar indicated the organisms were gram-negative. The number of bacteria pres-ent on the plate or how well they grew was not important.

However, there are many instances when it is critically important to know the number of bacteria present in a sample. A variety of methods can be used individually or in combination with other techniques to pro-vide an accurate measurement of viable cell numbers. One method involves using the colony counter to physically count each colony; another utilizes the spectrophotometer. A spectrophotometer is an instrument that can be used to determine bacterial cell numbers. A spectrophotometer measures the cell density or turbidity of a sample as a function of absorbance. In this exercise, the spectrophotometer will be used to measure culture turbidity as a quantitative but relative measurement of growth and cell numbers.

The theory behind spectrophotometry is simple. The machine directs a beam of light through a sample. Light either scatters, is absorbed, or passes through the sample **(figure 18.1)**. A sensor detects how much light passes through the specimen. Sensor data are presented numerically or graphically (needle scale) as either absorbance or transmittance. The more cells there are in the sample, the more light is absorbed by the cells and the less light passes through the sample.

Spectrophotometric measurements reflect either the light absorbed (absorbance, turbidity) by the sample or the light transmitted (transmittance) through a culture. Both of these measurements are proportional to the number of bacteria in a solution. The more turbid the culture, the more bacteria are present. Absorbance is a measurement of the amount of light absorbed or dispersed by the specimen. Absorbance is directly proportional to cell number. The more cells that are present in the sample, the higher the absorbance. Transmittance measures how much light passes through the sample. High transmittance values indicate that more light is passing through the sample, which also means there are fewer cells in the sample. Transmittance values are inversely proportional to cell number. In this exercise, you will measure absorbance of your cultures.

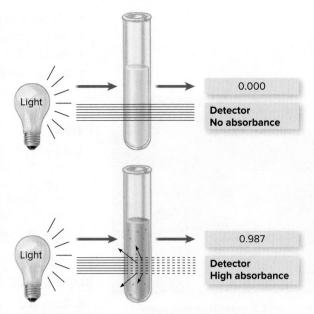

**Figure 18.1 Measuring absorbance with a spectrophotometer.** In the upper image, there are no cells in the sample (blank). The absorbance value is zero. In the lower image, a large number of cells are present. They absorb a large amount of light, which translates to a large absorbance value. The amount of light that passes through the sample is measured using a detector that displays this value in units of absorbance.

## ✓ Tips for Success

- As you pick up the media tubes, make sure to **label each tube** with the initials of the media (yeast extract broth [YEB], nutrient broth [NB], glucose salts broth [GSB], or inorganic synthetic broth [ISB]). The media look similar, so label them now, not later.

- When you label the tubes, write on the glass just below the cap but above the liquid media. If the writing is below the liquid, it will affect your spectrophotometric results.

- Make sure to auto **zero the spectrophotometer with the appropriate media blank** when measuring the growth in a specific medium. For example, when you measure the absorbance of growth in yeast extract broth, make sure to zero the spectrophotometer with the YEB blank.

- Make sure to mix your cultures thoroughly before dispensing into cuvettes.

## Organisms

- *Escherichia coli*
- *Streptococcus sanguinis* (formerly known as *Streptococcus sanguis*)

## Materials (per Team of Four)

- 2 tubes of yeast extract broth (YEB)
- 2 tubes of nutrient broth (NB)
- 2 tubes of glucose salts broth (GSB)
- 2 tubes of inorganic synthetic broth (ISB)
- 1 mL pipette

- Pipette pump
- Spectrophotometer (period 2)
- Media blanks (period 2)
- Bunsen burner and striker

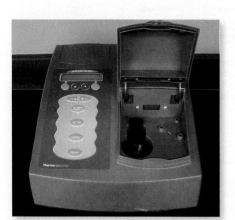

**Figure 18.2 Spectrophotometer.** The spectrophotometer is an instrument that measures the absorption of light by a sample.
©Steven D. Obenauf

## Procedure

### Period 1

1. Aseptically remove 0.1 mL (or 100 uL) of each organism and add to each tube. For example, 0.1 mL of *E. coli* will be added to one tube of YEB, NB, GSB, and ISB. Pipette accurately to ensure you are starting with the same number of cells in each aliquot. Repeat this procedure with *S. sanguinis*.

2. Place all eight tubes into the incubator.

### Period 2 (If you are not using a spectrophotometer, skip to step 6.)

1. Read and record the results of each organism in each medium using the **spectrophotometer (figure 18.2).** Depending on the type of spectrophotometer in the laboratory, you may read your results directly in your culture tubes or you may have to decant your cultures into cuvettes that fit into your spectrophotometer. Your instructor will provide specific directions. Your spectrophotometer should be set to read absorbance.

2. Zero the spectrophotometer using a blank (an uninoculated tube containing no cells). Each different medium has a separate blank. Pick up an uninoculated tube of glucose salts broth. **Resuspend the contents of the tube by carefully shaking or agitating the tube.** Wipe the outside of the tube (cuvette) carefully to remove any fingerprints or marks. Insert the tube into the spectrophotometer. Close the cover and press the button to zero the machine. Once the readings have stabilized, open the lid and remove the blank.

3. Retrieve the tube of GSB inoculated with *E. coli.* Roll the tube between your hands to resuspend the culture. Wipe the outside of the tube, insert it into the spectrophotometer, and close the lid. Record your results. Repeat this step with your culture of *S. sanguinis* growing in GSB.

4. Repeat steps 2 and 3 for YEB, ISB, and NB cultures. Make sure you re-blank the spectrophotometer with the proper blank.

5. Record your results in your laboratory report as absorbance units.

### Qualitative observation

6. Group your tubes in a series, so that you have all of your tubes inoculated with *E. coli* in one area and all of the tubes inoculated with *S. sanguinis* in another.

7. Resuspend each tube by gently agitating the tube. Compare the growth in each series.

8. If no growth is evident, record a "0" in the laboratory report. If a small amount of growth is present, record a "+." Record "++" for a moderate amount of growth and either "+++" or "++++" for more luxuriant growth.

9. Repeat this procedure for the other series of tubes.

# Chemical Content of Media

## Your Results and Observations

Record your results. Indicate in the far right column if the medium is classified as complex or defined.

| Organism or sample | Absorbance: *Escherichia coli* | Absorbance: *Streptococcus sanguinis* | Complex medium or defined medium |
|---|---|---|---|
| Yeast extract broth (YEB) | | | |
| Nutrient broth (NB) | | | |
| Glucose salts broth (GSB) | | | |
| Inorganic synthetic broth (ISB) | | | |

## Interpretation and Questions

1. Based on these recipes, what type of medium (complex or defined) is (a) nutrient broth or (b) glucose salts broth?

(a) _____  (b) _____

| Nutrient broth | | Glucose salts broth | |
|---|---|---|---|
| Peptone | 5 g | NaCl | 5.0 g |
| Beef extract | 3 g | $MgSO_4$ | 0.02 g |
| Water | Add to 1 L | $NH_4H_2PO_4$ | 1.0 g |
| | | $K_2HPO_4$ | 1.0 g |
| | | Glucose | 5.0 g |
| | | Water | Add to 1 L |

**2.** Based on your results, which medium supported the best overall growth? Is this medium a defined or an undefined medium?

_____

_____

_____

_____

**3.** Based on your results, which organism is more fastidious, _Escherichia coli or Streptococcus sanguinis_? Explain your answer.

_____

_____

_____

_____

**4.** Consider the evolution of human pathogenic bacteria. Would human pathogens be classified as fastidious or not? Explain your answer.

_____

_____

_____

_____

# Microbial Nutrition, Ecology, and Growth

# Osmotic Pressure and Growth

### CASE FILE

In mid-June, a nurse practitioner is evaluating a 61-year-old patient at a walk-in urgent care center. The patient's daughter, who brought him, reports that 9 days ago he began to suffer from abdominal discomfort, diarrhea, vomiting, and fever. Over the last day his mental state has changed and he has developed pain in his legs and widespread spots on his skin. On examination he is found to have significantly low blood pressure. Further investigation reveals that his gastrointestinal symptoms began a day after he ate some uncooked oysters. It is also determined he was diagnosed with hepatitis C 2 years ago.

As the history is suggestive of infection with *Vibrio vulnificus,* the patient is transported to a hospital, where immediate Gram staining and culture of his blood are ordered, and treatment with antibiotics and agents to raise his blood pressure is initiated. *V. vulnificus* lives in coastal saltwater areas and causes an infection, which often starts after eating seafood, especially raw or undercooked oysters. This organism can spread into the bloodstream in people with compromised immune systems, especially those with chronic liver disease.

### LEARNING OUTCOMES

At the completion of this exercise, students should be able to

- understand the effect of different concentrations of salt on bacterial growth.
- be able to interpret growth data and use them to categorize the salt preferences of bacteria being tested.

## Background

The cytoplasm of a cell is water based and is the site of most of the important cellular functions within a bacterium. Therefore, the ability of a bacterium to control the movement of water into and out of its cell is critical for its survival. Water movement across a semipermeable membrane is termed **osmosis.** Osmosis is regulated not by water but by the concentration of solutes contained within that water. *Solute* is a generic term for a compound that dissolves in a solution. In living cells, solutes include proteins, carbohydrates, and salts or ions. Dissolved solutes form hydration layers in which water coats the solutes through electrostatic interactions. This effectively lowers the concentration of "free" or "available" water.

In order to predict the movement of water, one must compare the salinity of the cell's cytoplasm to the salinity or solute concentration of the environment around the cell **(figure 19.1).** If the concentration of solutes outside the cell is *higher* than inside the cell, the solution surrounding the cell is considered **hypertonic.** In a hypertonic solution, water moves out of the cell. The loss of water causes the volume of the cytoplasm to decrease, and the plasma membrane "shrinks" around the remaining cytoplasm. This shrinking is called **plasmolysis.** If the concentration of solutes outside the cell is *lower* than inside the cell, the solution surrounding the cell is considered **hypotonic.** Under hypotonic conditions, water moves across the cytoplasmic membrane to the area of higher solute content (lower concentration of free water) until equilibrium is achieved. A bacterial cell bathed in a hypotonic environment will have increased internal pressure due to the influx of water into the cell.

Bacteria require turgor pressure to sustain life. **Turgor pressure** is created by the influx of water into a bacterial cell. Unlike animal cells, which tend to burst in hypotonic environments when too much water enters the cell, the expansion of a bacterial cell is limited by its peptidoglycan cell wall. The amount of pressure exerted by the cell against the cell wall reaches equilibrium with the pressure exerted by the cell wall against the cell. Bacteria tend to store or import ions in order to maintain turgor pressure.

Conversely, a bacterial cell must protect against the loss of too much water from the cytoplasm. Excessive water loss causes the plasma membrane to pull away from the cell wall (figure 19.1). This is called **plasmolysis.** Plasmolysis makes it difficult for a bacterial cell to carry out essential functions required for growth.

Most bacteria have an internal salinity equivalent to 0.9% salt (NaCl) and can grow in an environment that is equivalent to 3% NaCl. At concentrations over 3%, susceptible cells cannot control the loss of water, resulting in plasmolysis. However, some bacteria, called **halophiles,** can grow at higher salt concentrations due to their ability to maintain high enough cytoplasmic concentrations of solutes, preventing the excessive loss of water. **Facultative halophile** is a term used to describe bacteria that grow in both low-salt and high-salt (greater than 3% NaCl) environments. An **obligate halophile** is a bacterium that can grow only in the presence of high salt. *Halobacterium* is an example of an obligate halophile.

Most bacteria that cause disease in humans cannot grow in high-salt concentrations. Two bacteria that can are *Vibrio* and *Staphylococcus.*

## ✓ Tips for Success

- Make sure to label the different broths as you pick them up.

## Organisms (Broth Cultures)

- *Escherichia coli*
- *Staphylococcus aureus*

## Materials (per Team of Four)

- 2 1% NaCl broths
- 2 3% NaCl broths
- 2 7% NaCl broths
- 2 11% NaCl broths
- Bunsen burner and striker
- Loop

©Steven D. Obenauf

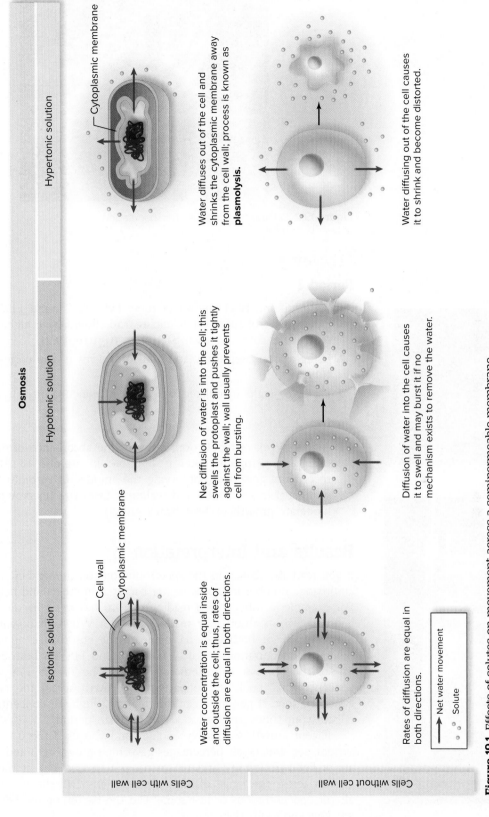

**Osmosis**

Isotonic solution | Hypotonic solution | Hypertonic solution

**Cells with cell wall**

Cell wall
Cytoplasmic membrane

Cytoplasmic membrane

Water concentration is equal inside and outside the cell; thus, rates of diffusion are equal in both directions.

Net diffusion of water is into the cell; this swells the protoplast and pushes it tightly against the wall; wall usually prevents cell from bursting.

Water diffuses out of the cell and shrinks the cytoplasmic membrane away from the cell wall; process is known as **plasmolysis.**

**Cells without cell wall**

Rates of diffusion are equal in both directions.

Diffusion of water into the cell causes it to swell and may burst it if no mechanism exists to remove the water.

Water diffusing out of the cell causes it to shrink and become distorted.

→ Net water movement

∘ Solute

**Figure 19.1** Effects of solutes on movement across a semipermeable membrane.

129

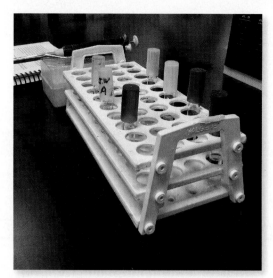

**Figure 19.2** Setting up the tubes for inoculation.
©Steven D. Obenauf

## Procedure

### Period 1

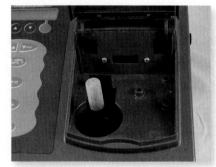

©Steven D. Obenauf

1. Set aside and label one set of tubes (1%, 3%, 7%, and 11%) for *E. coli* and one set of tubes for *S. aureus*. You should have eight tubes: a set of four tubes for each bacterium **(figure 19.2).**

2. Using a loop, aseptically add your organisms to the corresponding tube.

3. Place the eight inoculated tubes into the incubator.

### Period 2

1. After a week of incubation, measure the absorbance of each tube using a spectrophotometer as described in exercise 18. Alternatively, record the growth by observing the amount of turbidity (and sediment) in each tube and scaling it from 0 = no growth to +2 = moderate growth to +4 = heavy growth.

## Results and Interpretation

©Steven D. Obenauf

If you read absorbance in the spectrophotometer, remember that a higher absorbance number indicates higher bacterial numbers and better growth. For example, a tube with a reading of .540 has more bacteria in it than a tube with an absorbance reading of .045. When all of your numbers are recorded for both bacteria, compare the absorbance numbers at the different concentrations of salt (1%, 3%, 7%, and 11%) for both bacteria. The 1% salt tube is close to isotonic conditions. The 7% and 11% salt are high concentrations. Look at the pattern of growth across the range of salt concentrations. If neither organism grew at a high-salt concentration, reincubate your tubes, and reread them next lab period.

If your growth was heavy enough, and you used the 0 to +4 reading system, see which salt percentages gave high growth (+3 and +4), moderate or light growth (+2 and +1), or no growth (0). Look at the pattern of growth across the range of salt concentrations. Did the bacteria grow at all salt concentrations or just some? Did the bacteria grow better at some salt concentrations than others?

Name _____

Date _____

# Osmotic Pressure and Growth

## Your Results and Observations

Record the amount of growth (0 to +4) or the absorbance value for each organism in each salt concentration on both the table and graph that follow.

| Organism | 1% NaCl | 3% NaCl | 7% NaCl | 11% NaCl |
|---|---|---|---|---|
| *Escherichia coli* | | | | |
| *Staphylococcus aureus* | | | | |

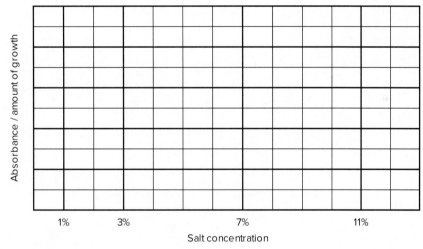

*E. coli* ..............

*S. aureus* _ _ _ _ _

## Interpretation and Questions

**1.** Which organism is considered a halophile? Explain.

_____

_____

_____

_____

_____

**2.** Is this organism an obligate halophile or a facultative halophile? Explain.

_____

_____

_____

_____

# Microbial Nutrition, Ecology, and Growth

# Oxygen and Growth: Aerotolerance

## CASE FILE

A 77-year-old male resident of southern Colorado was brought to the emergency room somewhat disoriented and with swelling and severe pain in his leg. The triage nurse observed that the patient's skin was discolored and blistering and there was an unpleasant odor. The patient's history revealed that he had diabetes, which was poorly controlled with metformin and had injured his leg in a fall 2 months previously. An X ray revealed pockets of gas between tissue layers in the lower leg. Blood and tissue samples were sent for culture. The patient was moved to the ICU and treatment was immediately begun with piperacillin, moxifloxacin, and vancomycin. The patient's family was counseled that depending on the results of the culture and the progression of his disease, he might need amputation or treatment in a hyperbaric chamber.

©kaling2100/Shutterstock. com RF

### LEARNING OUTCOMES

At the completion of this exercise, students should be able to

- describe the various ways bacteria carry out metabolism in the absence of oxygen.
- interpret growth patterns in brain-heart infusion (BHI) agar deeps and determine what category of aerotolerance they represent.

## Background

Bacteria are unique organisms in that they can inhabit virtually unlimited ecological niches. This is because bacteria possess diverse metabolic pathways. One growth requirement that varies dramatically among different bacteria is the utilization of molecular oxygen ($O_2$). Under aerobic conditions, oxygen acts as the final electron acceptor in the electron transport chain located in the plasma membrane of bacteria as part of a process termed **aerobic cellular respiration.** The shuttling of electrons through the electron transport chain to $O_2$ allows for the establishment of an electrochemical gradient differences in hydrogen ions ($H^+$) and charge across the plasma membrane, which eventually result in the production of adenosine triphosphate (ATP) via the membrane-bound enzyme ATP synthase. In the absence of oxygen, some bacteria shift their metabolic pathways to include anaerobic respiration and/or fermentation. In the process of

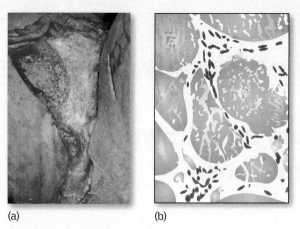

**Figure 20.1** Images from infections caused by an anaerobe. **(a)** A patient with gas gangrene. **(b)** Stained tissue shows pockets of gas caused by *Clostridium.*
©Michael English, M.D./Custom Medical Stock Photo/Newscom

**anaerobic respiration,** alternatives to $O_2$ are utilized as the final electron acceptor in the electron transport chain. These alternatives include nitrate ($NO_3^-$), sulfate ($SO_4^{2-}$), sulfur (S), and carbonate ($CO_3^{2-}$). In **fermentation,** organic molecules like pyruvate and pyruvate derivatives act as the final electron acceptors, allowing glycolysis to continue in the absence of oxygen. Each round of glycolysis results in the net production of 2 ATPs per glucose.

Organisms that solely rely on oxygen for cellular respiration are termed **obligate aerobes.** An organism that can use oxygen for ATP production but can also grow without it is termed a **facultative anaerobe.** An **obligate anaerobe** is an organism that is unable to grow in the presence of oxygen due to the toxic nature of alternative forms of oxygen. Bacteria that don't use oxygen to produce ATP but can tolerate (grow) in the presence of oxygen are termed **aerotolerant anaerobes. Microaerophiles** are bacteria that use oxygen for energy production, but only in very low concentrations. High levels of oxygen are toxic for microaerophiles. The ability to grow in the presence or absence of oxygen allows bacteria to cause infections throughout the body, including deeper tissues or tissues with poor blood supply and **poor oxygenation.** Facultative bacteria such as *Escherichia coli* and *Enterococcus* and anaerobes such as *Bacteroides, Fusobacterium,* and *Clostridium* can all cause disease. Anaerobes can often be challenging to isolate from tissues and grow in the lab. One particularly dangerous infection of deeper tissues is **gas gangrene** (also known as *clostridial myonecrosis*), caused by the anaerobe *Clostridium perfringens.* Tissue death and swelling caused by gas in the tissues are often seen in this deadly infection **(figure 20.1*a, b*).**

Culturing and examining bacteria with different oxygen requirements can be done in a number of ways, including using thioglycollate medium, special chambers, and BHI agar. This exercise utilizes **brain-heart infusion (BHI) agar.** BHI medium is an enriched medium for the cultivation of fastidious microorganisms. The agar is heated immediately before use to drive off soluble oxygen in the medium. The tempered agar is then inoculated and immediately cooled in an ice bath. The agar solidifies. Oxygen diffusion into the medium is now limited. This creates a gradient from high oxygen levels at the surface of the BHI deep tube to low oxygen levels at the bottom of the tube.

The growth patterns of different organisms in BHI agar based on their oxygen metabolism are (1) **obligate aerobes:** growth only at the top surface of the agar; (2) **facultative anaerobes:** growth throughout the agar and on the top; (3) **obligate anaerobes:** growth only at the bottom of the tube; (4) **aerotolerant anaerobes:** growth below the surface of the agar but not on it; and (5) **microaerophiles:** heaviest growth in a band just below the surface of the agar where the oxygen concentration is optimal.

## ✓ Tips for Success

- Take one tube out of the water bath at a time. Allow the tube to cool for a few minutes to avoid killing your cultures. However, if the tubes sit out of the bath for too long, the media will solidify before adding the inoculum.
- Make sure to **roll the tube** in your hands after adding the inoculum to evenly distribute the culture within the medium.
- If you put the bacteria on ice, don't leave the bacteria there. Remember to place the tubes in the incubator for growth.

## Organisms (Broth Cultures)

- *Escherichia coli*
- *Clostridium sporogenes* (available in GasPak anaerobic jar or pouch)
- *Alcaligenes faecalis, Micrococcus,* or *Pseudomonas aeruginosa*
- *Streptococcus sanguinis, Campylobacter jejuni, Enterococcus faecalis,* or *Bacillus megaterium*

## Materials (per Team of Two or Four)

- 4 melted BHI agar deeps in a 62°C bath
- Test tube rack
- 4 1 mL pipettes
- Pipette pump
- Bunsen burner and striker
- Ice bath

## Procedure

### Period 1

1. Remove a single BHI deep from the water bath. Take only one tube at a time!
2. Aseptically remove 0.1 mL of a bacterial culture using the 1.0 mL pipette. Three of the broth cultures will be in their normal locations; the fourth, *Clostridium,* may be in a GasPak jar **(figure 20.2)** or other container for anaerobic growth.
3. Insert the pipette to the bottom of the BHI tube and slowly dispense the inoculum while raising the pipette through the medium. Alternatively, insert a heavy loopful of bacteria all the way to the bottom of the tube.
4. Roll the tube in your hand, mixing up the contents of the tube.
5. Leave the tubes out at room temperature or place the tubes in an ice bath until completely solidified.
6. Repeat for all organisms.
7. Remove the tubes from the ice bath and place in incubator.

### Period 2

1. Read and record the results of each organism.

**Figure 20.2** GasPak anaerobic jar containing bacterial cultures.
©Steven D. Obenauf

**(a)**                    **(b)**                    **(c)**                    **(d)**

**Figure 20.3 Patterns of growth. (a)** Strict aerobe; **(b)** aerotolerant anaerobe; **(c)** facultative anaerobe; **(d)** microaerophile. ©Steven D. Obenauf

## Results and Interpretation

Observe the tubes for growth both on top of the agar and inside of the agar from top to bottom. Growth on the top is often off-white, while growth inside the agar will show as cloudiness (uninoculated BHI is clear). You may use an uninoculated tube of BHI as a control if needed. Growth only at the surface of the medium indicates an obligate aerobe **(figure 20.3a).** No growth or very slight growth on the surface with growth below the surface of the medium all the way to the bottom and cracks due to $CO_2$ production indicate an aerotolerant anaerobe (see the line and arrow on **figure 20.3b).** Heavy growth on the surface and growth throughout the medium from top to bottom and cracks due to gas production indicate a facultative anaerobe **(figure 20.3c).** A band or area of heaviest growth just below the surface of the medium indicates a microaerophile (see the arrow on **figure 20.3d).** If you wish, do a Google or Bing image search for "gas gangrene" to see the impact the causative organism in the Case File for this exercise can have on the human body.

# Oxygen and Growth: Aerotolerance

## Your Results and Observations

Draw or photograph your results.

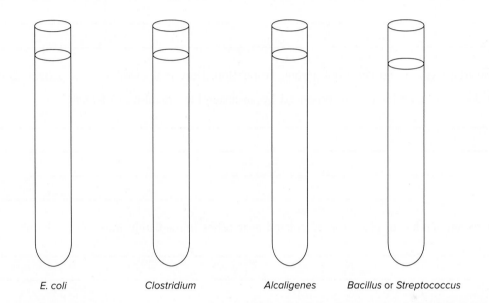

E. coli          Clostridium          Alcaligenes          Bacillus or Streptococcus

| Organism or sample | Description of the growth within the tube | How would you classify the organism based on its oxygen requirements. |
|---|---|---|
| Escherichia coli | | |
| Clostridium sporogenes | | |
| Alcaligenes faecalis or Micrococcus | | |
| Streptococcus or Bacillus megaterium | | |

## Interpretation and Questions

1. What causes the cracking in the agar? Can that same phenomenon occur in human tissue?

_____

_____

_____

_____

2. A sample from a patient with a suspected anaerobic infection is left out too long before being taken to the laboratory for growth and identification. Why would that be a problem?

_____

_____

_____

_____

3. Humans breathe oxygen and carry out aerobic respiration. How is it possible for a patient to contract tetanus, a disease caused by a toxin produced by an anaerobic organism, _C. tetani_?

_____

_____

_____

_____

4. Why is it recommended to place the inoculated agar tubes immediately into the ice bath?

_____

_____

_____

_____

## Microbial Nutrition, Ecology, and Growth

EXERCISE

# 21

# Oxygen and Growth: Catalase

## CASE FILE

Patient G is a 50-year-old woman. She developed several intraab-dominal abscesses following surgery for a kidney transplant. Treatment with IV ampicillin and gentamicin was not effective. The patient's temperature and white blood cell count both continued to be elevated. After an abdominal CT scan, the abscesses were drained and samples were sent for testing. Gram stains and blood cultures were performed at the same time, and both came back positive for gram-positive cocci. Further test-ing is under way to identify the organism. Treatment with vancomycin has been initiated pending the results of the further testing. A consulta-tion with a clinical pharmacist was requested.

©Chris Ryan/Getty Images RF

### LEARNING OUTCOMES

At the completion of this exercise, students should be able to

• understand the function of catalase in cells that produce the enzyme.

• perform and interpret the results of a catalase test and know their value in differentiating bacteria.

## Background

As the bacterial cell carries out different metabolic pathways, a harmful by-product called **hydrogen peroxide ($H_2O_2$)** is produced. Hydrogen peroxide is produced by the electron transport chain during the reduction of oxygen. Oxygen acts as the final electron acceptor in the process of aerobic respiration. Enzymes of the electron transport chain called oxidases reduce oxygen to form, among other products, hydrogen peroxide ($H_2O_2$). $H_2O_2$ is considered a **reactive oxygen species (ROS)** and can cause significant damage to important molecules of the bacterium like DNA, RNA, and proteins. To detoxify this ROS, many bacteria produce the enzyme **catalase.** The function of catalase is to convert $H_2O_2$ into $H_2O$ and $O_2$ gas, which are harmless to the cell.

$$2\,H_2O_2 \rightarrow 2\,H_2O + O_2$$

By detoxifying $H_2O_2$, catalase allows the bacterium to survive and thrive in oxygen-rich environments.

To determine if an organism produces catalase, an inoculum is placed on a glass slide. A few drops of $H_2O_2$ are then dispensed onto the cells.

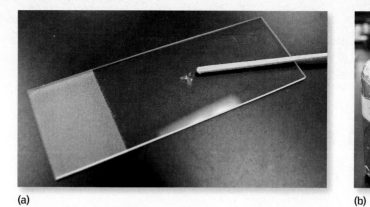

(a)      (b)

**Figure 21.1 Materials for performing the catalase test. (a)** Using a toothpick to remove a sample and adding it to the slide. **(b)** A 3% hydrogen peroxide solution. ©Steven D. Obenauf

If the organism produces catalase, bubbling will appear due to the production of $O_2$ gas from the breakdown of $H_2O_2$ and the organism is considered catalase-positive.

This test is useful in distinguishing gram-positive cocci from one another. *Staphylococcus* and *Micrococcus*, which are both catalase-positive, are differentiated from *Enterococcus* and *Streptococcus*, which are both catalase-negative. Both streptococcal and staphylococcal infections can sometimes be difficult to treat. *Enterococcus* is of great concern as a cause of hospital-acquired infections due to its natural resistance to multiple antibiotics, including penicillins, cephalosporins, and sulfamethoxazole, and its more recent acquisition of resistance to vancomycin (VRE) or fluoro-quinolones. Treating hospital-acquired *Enterococcus* often requires newer antibiotics or combinations of antibiotics.

## ✔ Tips for Success

- Make sure you have enough bacteria on your toothpick. If you use too small of a sample, you can get a false-negative reaction.

## Organisms (Slants or Plates)

- Patient G culture for diagnosis
- *Staphylococcus aureus* or *Micrococcus luteus*
- *Enterococcus faecalis*

## Materials (per Team of Two or Four)

- 1 clean microscope slide
- 3% $H_2O_2$ solution (store in the cold and protect from light)
- Toothpicks or swabs
- Sharps container

## Procedure

1. Scrape some cells of the colony (the amount should be enough that it is clearly visible on the end of your loop or toothpick) off of the surface of your plate or slant with a loop or toothpick.
2. Add the cells from your plate or slant to the glass slide with your loop or the end of a loop or toothpick, as shown in **figure 21.1a**.
3. Place one or two drops of $H_2O_2$ **(figure 21.1b)** onto the cells.

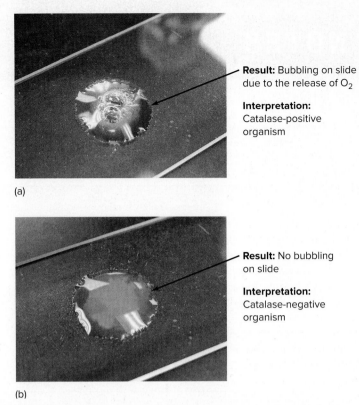

**Result:** Bubbling on slide due to the release of $O_2$

**Interpretation:**
Catalase-positive organism

(a)

**Result:** No bubbling on slide

**Interpretation:**
Catalase-negative organism

(b)

**Figure 21.2 Catalase test reactions. (a)** Bubbling due to the production of oxygen from the breakdown of $H_2O_2$; **(b)** no reaction of $H_2O_2$.
©Steven D. Obenauf

4. Watch for bubbling as an indication of $O_2$ production via the activities of the enzyme catalase.

5. Discard the used slide in a sharps container or other location designated by your instructor after you have recorded your results.

## Results and Interpretation

Bubbles should appear shortly after placing the $H_2O_2$ on the slide **(figure 21.2a).** Alternatively, catalase can be tested for by putting $H_2O_2$ on colonies on the surface of a plate. Depending on the strength of the reaction, the bubbles may be tiny or relatively large. *Any* bubbles indicate a positive reaction (figure 21.2a). A complete lack of bubbles indicates a negative reaction **(figure 21.2b).**

# NOTES

Name _____

Date _____

# Oxygen and Growth: Catalase

## Your Results and Observations

Record the reaction seen after adding hydrogen peroxide and if that result indicates catalase production.

| Organism or sample | Description of result | Production of catalase (+ or -) |
|---|---|---|
| Patient G culture | | |
| *Staphylococcus aureus* or *Micrococcus luteus* | | |
| *Enterococcus faecalis* | | |

## Interpretation and Questions

**1.** What organism is Patient G apparently infected with?

_____

_____

_____

_____

_____

**2.** Is treating Patient G with vancomycin likely to work? Why or why not?

_____

_____

_____

_____

# NOTES

EXERCISE

# 22

# Oxidase

## CASE FILE

Based on what he had seen in a television show, the patient (Patient B) was trying to melt metal in a homemade primitive forge in his yard when he received second- and third-degree burns covering his chest and his arms. After transport to the hospital, his vital signs were unstable with lowered blood pressure and elevated heart rate. The patient was stabilized with medication and fluid supplementation by IV. Dressings and a topical antimicrobial (silver sulfadiazine) were applied. The patient was given a schedule of wound care appointments as well as range of motion therapy. A dietitian prescribed a high-protein, high-calorie diet. One of the nurses, who was new to the unit, was hesitant to remove the crusts over the patient's wounds during the dressing changes.

Source: CDC/Janice Haney Carr

Six days after the burn, his temperature elevated to 39°C. Significant redness appeared, especially at the edges of the wound on one arm, there was an exudate with a sweetish smell, and the crusts over the burn in that area darkened. One of the nurses changing his dressings observed they seemed to have a slight greenish coloration and noted that in the patient's records as well as informing other members of the care team. Swabs were taken from the area for staining, quantitative culture, and biochemical testing to identify the causative organism. Blood was also drawn for culture to diagnose possible sepsis.

A number of bacteria can cause burn wound infections. *Pseudomonas aeruginosa*, an oxidase-positive, gram-negative organism, and *Staphylococcus aureus* are the two most common ones. The clinical signs in this case are suggestive of *Pseudomonas*. Hospitals can harbor *Pseudomonas* that can serve as a source of infection. *Pseudomonas* has been found to contaminate the sinks and bed rails hospitals and in some cases has been cultured from the hands of nurses. *Pseudomonas* has also shown the ability to grow in a surprisingly wide range of materials, including iodophor antiseptics, quaternary ammonium compounds, and, in some strains, silver. Since *Pseudomonas* frequently exhibits antibiotic resistance, antibiotic-sensitivity testing (exercise 25) was immediately ordered.

LEARNING OUTCOMES

At the completion of this exercise, students should be able to

- explain the chemical basis for the oxidase test.
- interpret reactions in the oxidase test.

## Background

In the process of cellular respiration, high potential energy molecules like glucose are completely oxidized. These electrons are then shuttled to the electron transport chain located in the plasma membrane of the bacterium **(figure 22.1)**. The function of the cytochromes in the electron transport chain, through a series of redox reactions, is to transfer electrons and reduce $O_2$ in the last step. As these redox reactions are taking place, energy is released, and this is ultimately used to produce ATP. Cytochrome oxidase is the final enzyme of the electron transport chain that reduces $O_2$ to form $H_2O$. There are different forms of cytochrome oxidase, depending on the bacterium. One form is called **cytochrome *c* oxidase.** The purpose of this

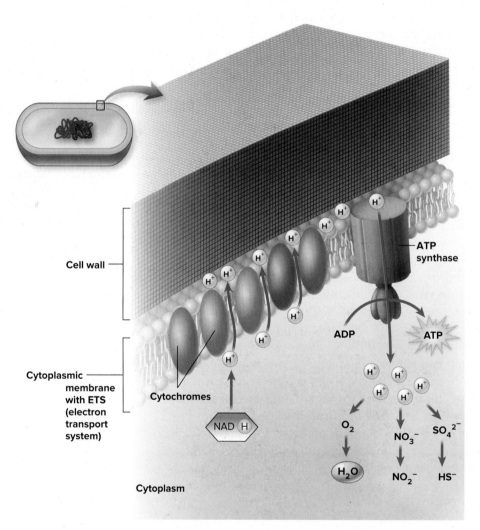

**Figure 22.1** The electron transport chain in bacteria.

test is to determine if the bacterium of interest has cytochrome *c* oxidase in its electron transport chain.

To test for the presence of this enzyme, a reagent called **oxidase reagent** is added to the cells of interest. The reagent is a redox indicator that is colorless when reduced and dark blue or purple when oxidized. The oxidase reagent will donate electrons to cytochrome *c* oxidase if it is present. By donating electrons, the reagent is oxidized and turns purple. This color change is an indication that the organism produces cytochrome *c* oxidase as part of its electron transport chain.

This procedure is used to distinguish among gram-negative rods. Members of the enterics, which are facultative anaerobes, do not utilize cytochrome *c* oxidase and are considered oxidase-negative. Aerobic, gram-negative *Pseudomonas aeruginosa* is oxidase-positive.

## ✓ Tips for Success

- When obtaining results, look for a blue-purple color change concentrated where the **cells** were added to the paper. As the reagent sits on the paper towel, it forms a blue ring, which DOES NOT indicate the presence of cytochrome *c* oxidase.
- This test must be done with cultures growing on solid media.

## Organisms (Slant or Plate Cultures)

- *Escherichia coli*
- *Pseudomonas aeruginosa*
- *Patient B sample*

## Materials (per Team of Two or Four)

- 1 paper towel or filter paper piece
- Toothpicks or loops
- Oxidase reagent dropper (BBL)

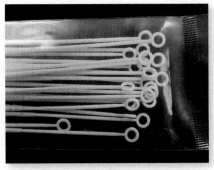
©Steven D. Obenauf

## Procedure

1. Draw two circles on your paper towel or filter paper, one for each organism to be tested.
2. Add the oxidase reagent (**figure 22.2**) to each circle. (Note: If you use a metal loop to add the organisms rather than plastic or wood, you will need to add the bacteria before the reagent to prevent false-positive reactions.)
3. Add each organism to a circle on the paper towel or filter paper using a plastic loop or toothpick. Use enough bacteria, so that they are visible on the paper. The reaction should start to develop within 1 minute. The area of the paper with the smeared cells (oxidase-positive) will start to turn blue and continue to get progressively darker.
4. Record the results.

## Results and Interpretation

The reaction will take place within a minute. Look at the color of the bacterial sample that was scraped onto the paper. If it turns blue-purple, that is a positive reaction (**figure 22.3, left).** A blue ring developing around the sample (figure 22.3, right) is not a positive reaction but is simply a reaction of the reagent with the paper.

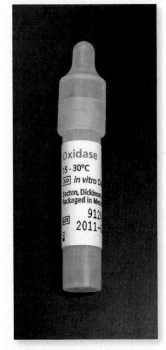

**Figure 22.2** Oxidase reagent vial.
©Steven D. Obenauf

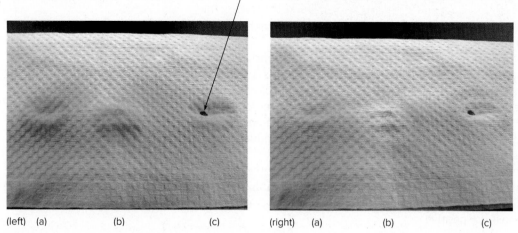

(left)  (a)          (b)          (c)          (right)  (a)          (b)          (c)

**Figure 22.3  Oxidase results.** The oxidase reagent has been added to organisms **a, b,** and **c. (left)** The reaction shortly after adding the reagent. **(right)** The reaction minutes later, after the paper has dried. Organism **c** produces cytochrome *c* oxidase as indicated by the blue color; organisms **a** and **b** do not produce cytochrome *c* oxidase. The faint blue-purple ring around each organism in the photograph on the right is caused by the reaction of the oxidase reagent with the paper. ©Steven D. Obenauf

# Oxidase

## Your Results and Observations

Record your results and whether they indicate a positive or negative reaction. Remember to record your color results within the first 2 minutes after the application of the oxidase reagent.

| Organism | Result (color of the bacterial sample) | Production of oxidase (+ or −) |
|---|---|---|
| *Escherichia coli* | | |
| *Pseudomonas aeruginosa* | | |

## Interpretation and Questions

**1.** You were distracted by another student in the classroom and forgot about your oxidase test for 15 minutes. When you check your test result, you notice a dark blue ring but no added coloration to the colony smear. What can you conclude from this result?

_____

_____

_____

_____

_____

**2.** Which is more likely to cause infections of deeper, less oxygenated tissues: a member of the enterics or *Pseudomonas*? Why?

_____

_____

_____

_____

_____

# NOTES

EXERCISE

23

# Plate Count

## CASE FILE

Bacteria are found everywhere. The challenge dairy farmers face is to produce milk with low numbers of bacteria in an environment that contains high numbers of bacteria (think about what you would find inside of a barn). Milk provides a perfect growth medium for microbes because of its water and nutrients. To reduce bacterial growth in milk, it must be cooled as quickly as possible after milking.

©Iconotec/Alamy Stock Photo RF

Bacteria reproduce by a process called binary fission, or the splitting of an organism into two separate organisms. This process can occur every 15 minutes under favorable conditions. This means that a tank of high-quality milk containing less than 1,000 bacteria per milliliter can change into poor-quality milk containing more than 1 million bacteria per milliliter in less than 5 hours if conditions are right. The bacteria in the tank include bacteria in milk from infected cows, bacteria normally present on teat skin, and bacteria found in dirt on the outside of the teats and udder. It also includes bacteria in dirt, water, and manure that may have entered the equipment. Improper cooling may also contribute to elevated bacterial counts because warm temperatures accelerate bacteria growth. Sampling is mandatory and federally regulated, and the bacterial count is determined using a standard testing procedure. Test results provide an estimate of the total number of bacteria present in the sample. The number, expressed as bacteria per milliliter, does not identify the bacteria present, just the estimated total number. FDA regulations require that before pasteurization, the count in samples containing mixed milk from many cows must be less than $3 \times 10^5$ organisms per milliliter. You are testing samples from a new dairy in your area that sells milk and ice cream at a creamery on the side of a highway to see if its milk is safe.

## LEARNING OUTCOMES

At the completion of this exercise, students should be able to
- perform a standard plate count.
- identify countable plates.
- calculate dilution factors and be able to calculate cells per milliliter given colony count and dilution factor.

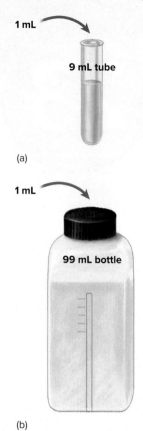

Figure 23.1 Single dilutions.
(a) Transferring 1 mL into a 9.0 mL tube gives a 1:10 ($10^{-1}$) dilution.
(b) Transferring 1 mL into a 99 mL bottle gives a 1:100 ($10^{-2}$) dilution.

# Background

The plate count, known as the heterotrophic plate count (HPC), standard plate count, or aerobic plate count, method is commonly used around the world for estimating the number of microbes in a sample. In this method, the original sample is diluted through a series of bottles or tubes. A set volume (aliquot) of the dilutions is then put onto plates, and the number of colonies on the plates is counted after incubation for 24 hours or more. The number of bacteria that were in the original sample can be calculated by multiplying the number of colonies on the plate by the dilution factor. Greater accuracy can be obtained by counting two or three plates and averaging the counts. In the standard method, only plates containing between 30 and 300 colonies are considered to be valid and countable. Plates on which fewer than 30 colonies are growing are labeled TFTC, too few to count. Plates on which more than 300 colonies are growing are labeled TNTC, too numerous to count. The assumption is that each colony arises from a single bacterial cell. Because this may not always be true, counts in the surface plate count method are reported as *colony-forming units,* or CFU. An advantage of the surface plate count method over directly counting microbes under the microscope is that both living and dead cells look the same under the microscope, but only living cells will grow on a plate. Plate counts, therefore, provide a better indicator of disease risk.

The standard plate count method is very widely used to estimate the number of bacteria in environmental samples, drinking water, and various juices and foods. This is used to determine the safety of where you swim and what you eat or drink. Safety standards from various federal or state agencies are usually expressed as a maximum allowable cells or CFU per milliliter or CFU per 100 mL.

## A Bit About Dilutions

Dilution of the sample is necessary to reduce the number in a sample to a level that can be counted. Although a lot of people try to avoid math as much as possible, an understanding of dilutions is important for calculating the amount of bacteria in the sample being tested and for knowing how to prepare a sample for testing.

Dilutions are all about ratios and how to express them. The formula for calculating a dilution can be stated as the volume of sample/(volume of sample + volume of diluent). If you add 1 mL of sample to a 9 mL tube of diluent, your dilution is 1/(1 + 9), or 1/10 **(figure 23.1).** The way to express 1/10 in scientific notation is $10^{-1}$. If you add 0.1 mL of sample to a 9.9 mL tube of diluent, your dilution is 0.1/(0.1 + 9.9), or 0.1/10, or 1/100 **(figure 23.2).** The way to express 1/100 in scientific notation is $10^{-2}$.

In serial dilutions (one dilution followed by another), you multiply all the dilutions used to get the final dilution of that sample **(figure 23.3).** If you make a 1/100 dilution in your first sample and then use that to make a 1:10 dilution into your second sample, the dilution in your second sample is (1/100) × (1/10), or 1/(10 × 100), or 1/1,000 (expressed as $10^{-3}$).

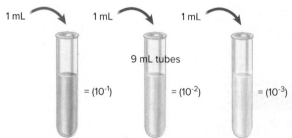

Figure 23.2 Serial dilutions: Transferring 1 mL into the first 9 mL tube gives a 1:10 ($10^{-1}$) dilution. Transferring 1 mL from the first 9 mL tube into the second one gives an additional 1:10 dilution to give a total dilution of 1:100 ($10^{-2}$). A third transfer would give a total dilution of 1:1,000 ($10^{-3}$).

## Calculating Cell Concentration

The reason for doing a standard plate count is to find out the concentration of bacteria in the sample being tested. The formula for calculating cells per milliliter (CFU) in the original sample is the number of colonies on the plate × 1/(final dilution). For example: If a plate with a $10^{-6}$ final dilution has 86 colonies on it, your cells per milliliter are 86 × (1/$10^{-6}$). Thus, 1/$10^{-6}$ is $10^{6}$, so your count is 86 × $10^{6}$. Expressing this correctly in scientific notation gives us 8.6 × $10^{7}$ cells per milliliter.

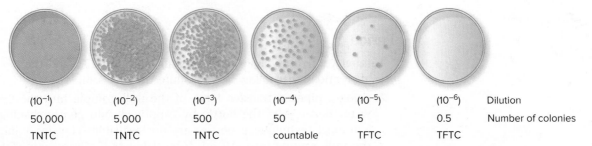

| (10⁻¹) | (10⁻²) | (10⁻³) | (10⁻⁴) | (10⁻⁵) | (10⁻⁶) | Dilution |
|---|---|---|---|---|---|---|
| 50,000 | 5,000 | 500 | 50 | 5 | 0.5 | Number of colonies |
| TNTC | TNTC | TNTC | countable | TFTC | TFTC | |

**Figure 23.3** Notice how colony numbers decrease as we dilute a sample containing 500,000 ($5 \times 10^5$) bacteria per milliliter.

## ✔ Tips for Success

- To avoid confusion, this lab works best if you divide it into three phases. First, make the dilutions (follow **figure 23.4**). Second, when all of the dilutions are done, pipette them onto the plates **(figure 23.5)**. Finally, when all the plates contain samples, spread them all.
- You will be using pipettes and pipette pumps and micropipettors in this experiment. Remember to maintain aseptic technique when handling pipettes and tips.
- The key to the success of this lab is making good dilutions. Pipette carefully.
- Use caution when flaming the spreader bar with alcohol.
- When spreading, make sure to cover the entire surface of the plate or your colonies may bunch together and be uncountable.

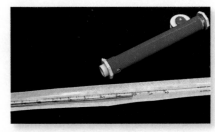

©Steven D. Obenauf

©Steven D. Obenauf

## Sample (per Team of Two or Four)

- Sample D (milk sample from dairy)

## Materials

- 8 nutrient agar plates
- 1 99 mL bottle of sterile water
- 3 9.0 mL tubes of sterile water
- Spreader bars
- Beaker containing alcohol
- Pipettes
- Pipette pumps
- Micropipettor
- Micropipettor tips
- Quebec Colony Counter (period 2)
- Bunsen burner and striker

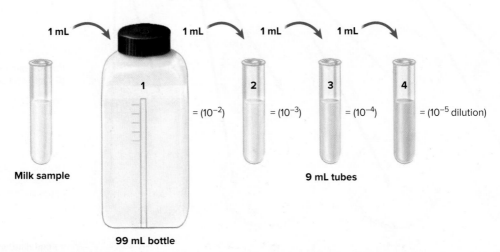

**Figure 23.4 Diluting the milk sample.** This schematic shows the steps involved in diluting a milk sample for plating. Each curved arrow represents a pipetting step. The numbers in parentheses indicate the total dilution at that step.

## Procedure

### Period 1

1. Label your tubes. Follow the diagram in figure 23.4 and the instructions that follow as you dilute the test samples.

2. Using a pipette, transfer 1 mL of the milk sample into the 99 mL water bottle. Mix the bottle by capping it and gently shaking it. (Optional technique if bottles are not available: Transfer 1 mL of the milk sample into a 9 mL tube (label as "1a") and mix. Transfer 1 mL from that tube into a second 9 mL tube (label as "1b") and mix.)

3. Using a new pipette, transfer 1 mL from bottle "1" to the first 9 mL tube ("2"). Mix the tube by rolling it between your hands. Do not shake the tube!

4. Using a new pipette, transfer 1 mL from the first test tube (labeled "2" in the diagram) to the second 9 mL tube (labeled "3" in the diagram). Mix the tube by rolling it between your hands.

5. Using a new pipette, transfer 1 mL from the second tube to the third 9 mL tube (labeled "4" in the diagram). Mix the tube by rolling it between your hands.

6. Label your plates on the bottom (make sure you mark them as the **final dilution**) and lay them out as shown in figure 23.5.

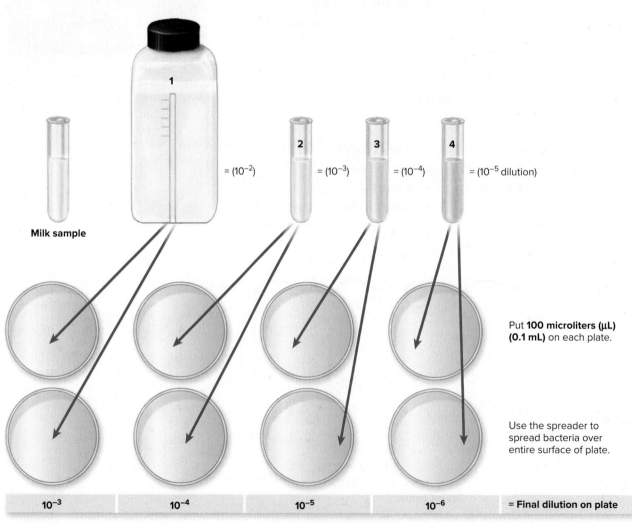

Milk sample

1

2 = $(10^{-2})$

3 = $(10^{-3})$

4 = $(10^{-4})$

= $(10^{-5}$ dilution)

Put **100 microliters (μL) (0.1 mL)** on each plate.

Use the spreader to spread bacteria over entire surface of plate.

| $10^{-3}$ | $10^{-4}$ | $10^{-5}$ | $10^{-6}$ | = Final dilution on plate |

**Figure 23.5** Delivering samples onto the plates and the resulting dilutions.

7. Pipette 100 µL (0.1 mL) of the diluted samples onto each of the plates with a micropipettor or pipette **(figure 23.6)**. Put the sample in the middle of each plate **(figure 23.7a)**. Change tips between samples.

8. Sterilize a spreader with alcohol using your Bunsen burner. Dip the spreader in the alcohol and light the alcohol by passing the spreader quickly through the flame of the Bunsen burner (do not leave the spreader in the burner flame or it may get too hot). **While flaming, make sure to keep the spreader pointing down and below your hand to prevent burns.**

9. Use the spreader bar to spread the samples over the entire surface of all of the plates **(figure 23.7b)**. Spread the samples as soon as possible after pipetting them onto the plates. It is best to have two students work together on this part of the exercise. If indicated by your professor, you may start spreading with the highest-dilution ($10^{-6}$) plates and then go to the $10^{-5}$, $10^{-4}$, and $10^{-3}$ plates in order. If you do this, you only need to flame the spreader before you start and after you finish, which is safer. If not, you need to flame between each dilution.

10. Dispose of used pipettes in a bucket or large sharps container. Dispose of used micropipettor tips in the sharps container. None of these items should go in the regular trash.

11. Place the plates lid side down in the incubator.

## Period 2

1. Lay out your plates ($10^{-3}$, $10^{-4}$, $10^{-5}$, and $10^{-6}$ dilutions) as in figure 23.5 and look at them.

2. Find the dilution with countable numbers of colonies (between 30 and 300) and count those plates using the Quebec Colony Counter **(figure 23.8).** Count all of the colonies on the plates. Alternatively,

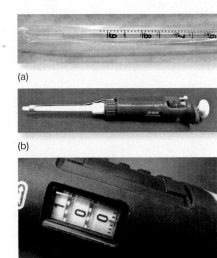

(a)

(b)

(c)

**Figure 23.6** Pipette options. You can deliver a volume of 100 microliters (0.1 mL) with either a regular pipette **(a)** or a micropipettor **(b and c)**. The micropipettor will be more accurate. (a-c) ©Steven D. Obenauf

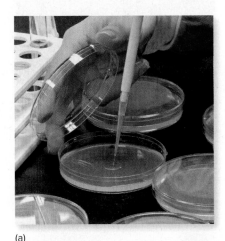

(a)

(b)

**Figure 23.7** Preparing plates. **(a)** Pipetting samples onto plates. **(b)** Spreading samples with spreader bar, sometimes called a "hockey stick." (a-b) ©Steven D. Obenauf

**Figure 23.8** Use the grid on the colony counter to "guide" you around the plate to make sure you count all the colonies. Some people use a marker and put a dot on the plate lid above each colony they count to prevent double counting or missing colonies. ©Steven D. Obenauf

you can turn the plates agar side up and count and mark each colony with a marker.

3. Record your results (under "cell/colony count") and the final dilution of the plates you counted (in all three boxes under "dilution of plates counted").

## Results and Interpretation

At lower dilutions, you may see plates with more than 300 colonies **(figure 23.9a)** on them (you can count this one if you like!), which you will record as TNTC. Plates with between 30 and 300 colonies **(figure 23.9b** has 82) are countable. Make sure you count every colony on the plate. At higher dilutions, you may see plates with less than 30 colonies on them **(figure 23.9c** has 11), which you will record as TFTC.

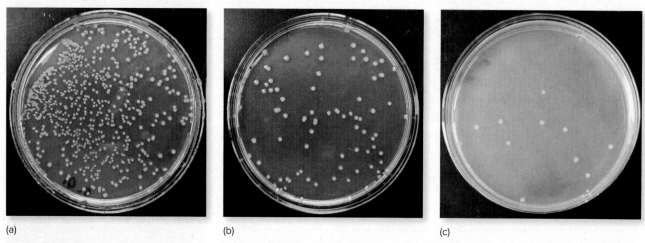

(a)                                              (b)                                              (c)

**Figure 23.9 Countable and uncountable plates. (a)** A TNTC plate with more than 300 colonies. **(b)** A countable plate. **(c)** A TFTC plate with less than 30 colonies. (a-c) ©Steven D. Obenauf

## Lab Report 23

# Plate Count

## Your Results and Observations

Draw or photograph with your mobile device and label (TFTC, TNTC) your plates.

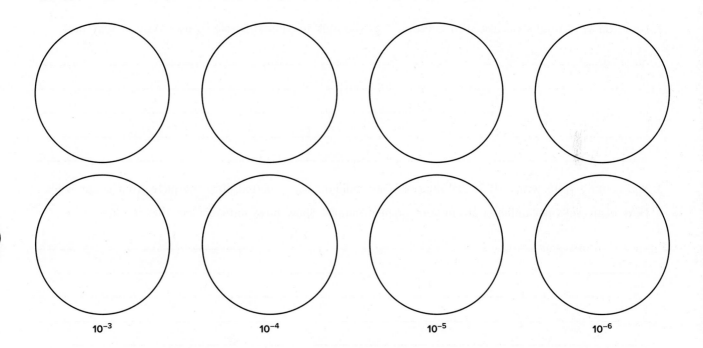

$10^{-3}$          $10^{-4}$          $10^{-5}$          $10^{-6}$

| Sample | Cell (colony) count | Dilution of plates counted |
|---|---|---|
| Plate 1 | | |
| Plate 2 | | |
| Average count of plate 1 and plate 2 | | |

# Interpretation and Questions

1. Based on the results recorded in the preceding table, how many cells per milliliter (CFU) are there in the milk sample? Express using scientific notation. The formula to use for your calculation is CFU = Average colony count × Final dilution of plate counted or average colony count × Dilution of plate counted × Volume placed on plate.

_____

_____

_____

_____

_____

2. Based on your results calculated in question 1, is the milk in Sample D safe to use? Why or why not?

_____

_____

_____

_____

_____

3. You count a plate with a $10^{-4}$ final dilution. One milliliter was pipetted onto the plate. Your count is 53. How many cells per milliliter are in your original sample? Show how you calculate that result.

_____

_____

_____

_____

_____

4. You count a plate with a $10^{-6}$ dilution. One-tenth milliliter was pipetted onto the plate. Your count is 129. How many cells per milliliter are in your original sample?

_____

_____

_____

_____

_____

5. You are testing a sample of water from a local swimming pool. Your $10^{-1}$ final dilution plate has 38 colonies on it. One milliliter was pipetted onto the plate. The safety standards for swimming pools state that the water cannot contain more than 200 CFU per milliliter (cells per milliliter). Is the pool you are testing safe to swim in? Why or why not?

_____

_____

_____

_____

_____

6. You transfer 1 mL of your test sample into a 99 mL bottle. After mixing, you transfer 1 mL from that first 99 mL bottle to a second 99 mL bottle. What is the dilution in that second bottle?

_____

_____

_____

_____

_____

7. As part of an ongoing monitoring program, you are testing a sample of water from the Chattahoochee River north of Atlanta. Your $10^0$ final dilution (undiluted) plate has 56 *Escherichia coli* colonies on it. One milliliter was pipetted onto the plate. The EPA safety standards for recreational freshwater state that the water cannot contain more than 235 *E. coli* per 100 mL. Is the river safe to swim or kayak in? Why or why not?

_____

_____

_____

_____

8. You are testing a sample of milk from a New York dairy that has been cited for some recent safety violations. Your $10^{-2}$ final dilution plate has 201 colonies on it. One milliliter was pipetted onto the plate. The FDA safety standard for milk of the type you sampled is $3 \times 10^5$ cells per milliliter. Is this milk sample safe to drink? Why or why not?

_____

_____

_____

_____

# NOTES

# Killing by Ultraviolet Light

## CASE FILE

Eight patients with severe respiratory disease presented at three different Cincinnati-area emergency rooms. Cough and a high fever (40.3°C [104.6°F]) were the most commonly seen symptoms. Respiratory samples from the patients yielded inconclusive results, but positive urinary tests led to the diagnosis of Legionnaire's disease. The patients all recovered after treatment with antibiotics. Interviews by the county board of health determined that all of the patients had used the whirlpool at a local spa. Samples of water from the spa yielded *Legionella pneumophila* of the same strain as the one infecting the patients. The spa was drained and disinfected. As *Legionella* is resistant to chlorine, the installation of an in-line ultraviolet (UV) disinfection system was recommended to prevent future outbreaks. UV light is a nonchemical alternative to the use of chlorine. Water typically flows through a special chamber housing a UV lamp with a wavelength between 240 and 280 nanometers. UV light, along with other alternatives such as ozone or microfiltration, is increasingly being used as an alternative to chlorine.

©Don Farrall/
Getty Images RF

## LEARNING OUTCOMES

At the completion of this exercise, students should be able to

- describe the mechanism by which UV light kills microorganisms.
- demonstrate the safety methods to be used when working near a UV light source.
- interpret the results of a UV light exposure procedure.

## Background

Methods for controlling or preventing the growth of microorganisms include chemical methods, physical methods, and antimicrobial drugs. One physical method that can be used to control microbial growth is radiation. The two forms of radiation that are used to kill microorganisms are gamma radiation and ultraviolet radiation. Gamma radiation (also known as ionizing radiation) has its greatest impact when it collides with molecules in the cell, creating free radicals, which are toxic. Ultraviolet radiation (also known as nonionizing radiation) primarily kills by damaging DNA. When

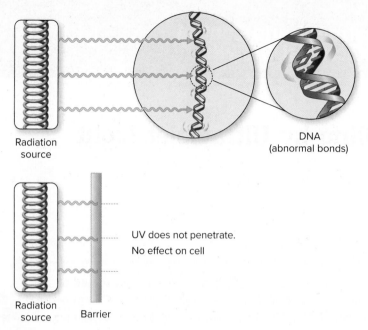

Radiation
source

DNA
(abnormal bonds)

UV does not penetrate.
No effect on cell

Radiation
source          Barrier

**Figure 24.1** Microbicidal effects of UV light.

DNA absorbs UV light, in particular UV-C, bonding relationships within the molecule can be affected. A major example is the formation of thymine or cytosine **dimers** (bonds between two adjacent pyrimidines in the DNA strand). These can lead to mutations **(figure 24.1).** UV light with a wavelength between 200 and 290 nanometers (nm) is the most lethal because these are the wavelengths that are absorbed the most by DNA. Exposure to a large amount of **UV-C** light (100–280 nm wavelength) will cause more damage than cell systems can repair, leading to death of the cell. You may have suffered from UV-B (280–315 nm wavelength) damage if you have spent too much time in the sun. However, the wavelengths that cause sunburn are not the same as those that kill microbes.

One of the biggest limitations that impact the use of UV light as a sterilizing method is its poor penetrating ability. Ultraviolet light will not go through plastic, glass, and many other materials. Because of this, it is mainly used for killing microbes on surfaces, in air, or in water. In addition, UV light is very damaging to the eyes and skin and so must be used when people are not present or exposed. The 254 nm UV-C light emitted by **germicidal lamps** is valuable for killing microbes in the air and on surfaces such as inside transfer hoods for tissue culture, clean rooms, and operating rooms.

## ✓ Tips for Success

- **Wear goggles** when working near the UV lamp. Avoid directly looking at the lamp (UV light can cause cataracts or other **eye damage**), and avoid exposing your hands to UV light. Even a few seconds of exposure to unprotected eyes can cause eye damage.

## Organisms (Plate Cultures)

- *Staphylococcus aureus* or *Serratia*
- *Bacillus cereus* or *Bacillus megaterium* (use a culture at least 1 week old; the older the culture, the better)

## Materials (per Team of Two or Four)

- Ultraviolet lamp (254 nm wavelength)
- 3 × 5 index cards or half-sheets of paper

©Steven D. Obenauf

162

- Box or shield to cover lamp
- 3 nutrient agar plates
- Inoculating loop or swab
- Bunsen burner and striker
- Gloves

## Procedure

### Period 1

1. Use your marker to divide each plate in half (mark the bottom of the plate as shown in **figure 24.2a**).
2. Label one side of the plate "covered" and the other side "uncovered." As an alternative, your instructor may have you make separate plates for "covered" and "uncovered." If you use separate plates, you will not need index cards.
3. Label one plate "A," one "B," and one "C."
4. Use a loop to streak each sample of bacteria on half of each plate (as shown in **figure 24.2b**).
5. Place plate A under the UV lamp **(figure 24.3a). Remove the lid** from the plate and cover half of the plate with an index card (as shown in **figure 24.3b**). Turn on the lamp and expose the plate for 30 seconds. Make sure you are wearing goggles and do not have your hands under the lamp.
6. Expose plate B for 1 minute. Expose plate C for 5 minutes.
7. Your professor may assign your group different times than the ones listed here.
8. Put the lids back on your plates and put the plates in the incubator lid side down.
9. The incubator must be dark (to prevent light repair of the DNA damage), and you should put the plates into the incubator as soon as possible.

### Period 2

1. Examine your plates for growth on each side.
2. Record your results.

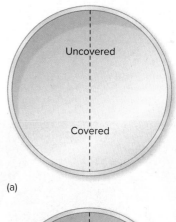

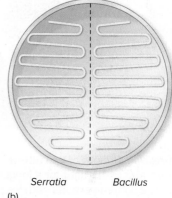

(a)

(b)

**Figure 24.2** Preparing the plates.
**(a)** Labeling a plate; **(b)** pattern for streaking the plates.

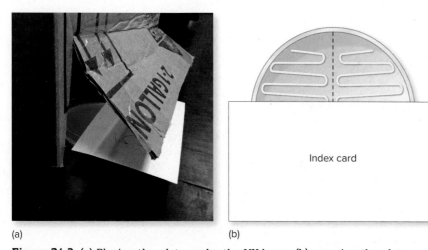

(a)

(b)

**Figure 24.3 (a)** Placing the plate under the UV lamp; **(b)** covering the plate with an index card. (a) ©Steven D. Obenauf

163

# Results and Interpretation

Your three plates may or may not look the same as one another and the plates shown here **(figure 24.4)**. Compare the growth on the covered side to the uncovered side (or covered and uncovered plates) at the various exposure times and compare the growth of *Bacillus* to that of *Serratia*.

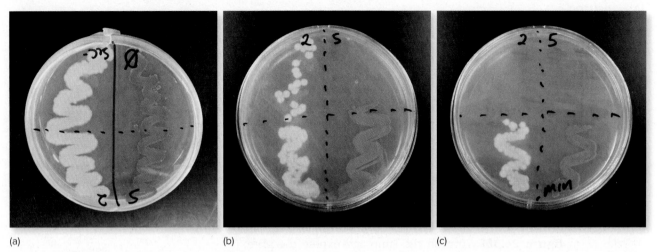

**Figure 24.4** Growth of two different bacteria on a plate exposed to **(a)** no UV light, **(b)** UV light for a moderate amount of time, and **(c)** UV light for a longer amount of time. The top half of the plates shown in **(b)** and **(c)** were exposed to UV light. The bottom portions of these plates were covered and not exposed to UV. (a-c) ©Steven D. Obenauf

Name _____

Date _____

# Killing by Ultraviolet Light

## Your Results and Observations

Draw or photograph your plates. Indicate the exposure time.

A: _____ minutes    B: _____ minutes    C: _____ minutes

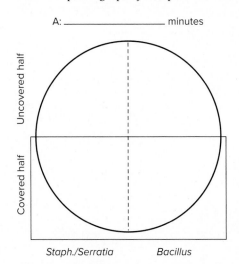

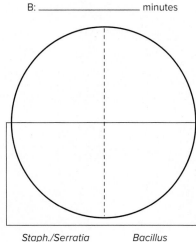

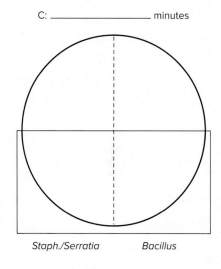

Staph./Serratia    Bacillus        Staph./Serratia    Bacillus        Staph./Serratia    Bacillus

## Interpretation and Questions

**1.** Based on your results, which time was the most effective? Least effective?

_____

_____

_____

_____

**2.** What difference would you expect to see in the results using an older *Bacillus* culture versus a younger one. Why?

_____

_____

_____

_____

**3.** UV light is used in clean rooms and some operating rooms. What are the limitations of using UV light as a means of sterilization?

_____

_____

_____

_____

**4.** If you were the quality control agent for a hospital using UV sterilization in an operating theater, how would you test its effectiveness in infection control?

_____

_____

_____

_____

# Antimicrobial Susceptibility Testing (Kirby-Bauer Method)

## CASE FILE

Antimicrobial susceptibility testing has been ordered for three patients. **Patient A,** a 4-year-old boy, was complaining that his throat hurt when his parents picked him up from day care. He was brought in for an examination. He was found to have a temperature of 39°C (102.2°F), a red throat, and some swelling of the lymph nodes in his neck. A whitish exudate was seen on his tonsils. No skin rash was visible. Diagnostic tests were ordered.

©Steven D. Obenauf

A throat swab was taken for culture on a blood agar plate (exercise 14) and a rapid immunoassay for group A *Streptococcus* (exercise 34). He was empirically started on amoxicillin. Because of previous problems while treating ear infections (otitis media), an antibiotic sensitivity test was ordered. **Patient B** is showing symptoms of infection following a burn. In addition to biochemical tests to identify the causative organism (exercise 23), an antibiotic sensitivity test was ordered, as the likely causative organism is difficult to treat. **Patient X** has developed a postsurgical infection. A multitest system is being used to identify the causative organism (exercise 34). As the infection may be hospital-acquired, tracing of the possible source of the infection is being done with DNA analysis (exercise 27).

### LEARNING OUTCOMES

At the completion of this exercise, students should be able to

- demonstrate how to perform antimicrobial susceptibility testing using the Kirby-Bauer method.
- understand the origin of a zone of inhibition and how it should be measured.
- interpret the results of the Kirby-Bauer method and determine if an antibiotic is effective against a particular bacterium.

## Background

*Antimicrobial* is a general term for something that kills or inhibits microorganisms. Antimicrobial drugs are one of the most widely used life-saving tools in medicine today. Very few drugs work equally well against all types of bacteria due to natural or acquired resistance on the part of the bacteria. Some, such as penicillin, are more effective against gram-positive bacteria, whereas others, such as Polymyxin B, have a greater effect against gram-negative bacteria. Azithromycin, cephalothin, and gentamicin kill some gram-negative and some gram-positive bacteria. Antibiotics with a broad spectrum of activity include ciprofloxacin, tetracycline, and trimethoprim. Amoxicillin/clavulanic acid (AmC) contains a compound to protect it against some drug-resistant bacteria. Clindamycin is sometimes used to treat methicillin-resistent *Staph. aureus* (MRSA). When treating an infected patient, we want to use the antibiotic that will be both the safest and the most effective against his or her disease. In many cases, we need to test the bacteria the patient is infected with against a variety of different antibiotics to see which will work the best. The goal is to find out whether the bacteria being tested are **sensitive** to the antibiotic (will be killed by it) or **resistant** to it (will be ineffective for treating the patient).

The **Kirby-Bauer method** is widely used for testing the sensitivity of bacteria to antimicrobials. The test is done by first covering the surface of a plate with the bacteria being tested. Paper discs containing known concentrations of different antimicrobials are then placed on the surface of the plate. The concentrations of the antimicrobial compound found in the discs are the same as the concentration found in the body of someone taking that antimicrobial. The antimicrobial agent in the disc diffuses out of it, forming a concentration gradient in the agar around the disc. The concentration of the antibiotic at the edge of the disc is high and gradually diminishes as the distance from the disc increases to a point where it no longer kills or inhibits the organism, which then grows freely. If the antimicrobial inhibits the growth of the organism, a clear, circular area appears around the disc. This is called the **zone of inhibition.** The method for doing the Kirby-Bauer test is highly standardized for use in clinical laboratories. It involves a special type of agar, **Mueller-Hinton agar,** which must be poured to a standard thickness. The plates are inoculated with cotton swabs rather than inoculating loops. Antibiotic discs **(figure 25.1)** are then placed on the surface of the plate either with tweezers or using a mechanical dispenser. After incubation, the plates are examined for zones of inhibition and the zones are measured. The measurements of the zones are then compared to standards to determine if the bacteria are sensitive to the antibiotic or resistant to it, or if the measurement is between those numbers (intermediate). Your patient probably won't be happy if you treat him or her with an antibiotic that has intermediate effectiveness or to which the bacteria are resistant!

## ✔ Tips for Success

- Swab the plates *before* adding the antibiotic discs.
- When you swab your plates, make sure to thoroughly swab the entire plate; **do not leave gaps**, or you may not see and measure the zones of inhibition well.
- Tap the center of each disc once you have placed it on the plate to make sure the disc is in contact with the agar. If not, you may have a false-negative result.
- Once the discs are all on the surface of the plates, keep them lid side down when you put them in the incubator. Water dropping onto the plate surface will ruin the zones of inhibition.

## Organisms (Broth Cultures)

- Patient X (gram-negative infection)
- Patient A (*Staphylococcus* or *Streptococcus* infection)
- Patient B (*Pseudomonas* infection)

## Materials

- 3 Mueller-Hinton plates
- Sterile swabs for inoculating the plates

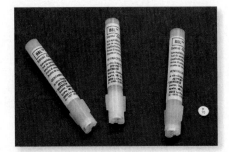

**Figure 25.1** BBL Sensi-Disc containers. ©Steven D. Obenauf

## Table 25.1 Measurement Standards*

| Antimicrobial | Zone of inhibition diameter (mm) | | |
| --- | --- | --- | --- |
| | Sensitive | Intermediate | Resistant |
| Amoxicillin AmC-30** | ≥ 18 mm | 14-17 mm | ≤ 13 mm |
| Azithromycin AZM-15 | ≥ 18 mm | 14-17 mm | ≤ 13 mm |
| Cephalothin CF-30 | zone ≥ 18 mm | zone 15-17 mm | zone ≤ 14 mm |
| Ciprofloxacin CIP-5 | ≥ 21 mm | 16-20 mm | ≤ 15 mm |
| Clindamycin CC-2 | ≥ 21 mm | 15-20 mm | ≤ 14 mm |
| Gentamicin GM-10 | ≥ 15 mm | 13-14 mm | ≤ 12 mm |
| Penicillin P-10 | ≥ 29 mm | | ≤ 28 mm |
| Polymyxin B PB300 | ≥ 12 mm | 9-11 mm | ≤ 8 mm |
| Tetracycline TE-30 | ≥ 19 mm | 15-18 mm | ≤ 14 mm |
| Trimethoprim TMP-5 | ≥ 16 mm | 11-15 mm | ≤ 10 mm |

*These values are the standard clinical values used in measuring bacterial antibiotic sensitivity.

**For *Staphylococcus*, sensitive is ≥ 29 mm; resistant is ≤ 9 mm with amoxicillin/clavulanic acid.

- Antibiotic discs (BBL Sensi-Discs): ciprofloxacin (5 micrograms [μg]), gentamicin (10 μg), penicillin (10 units), azithromycin (15 μg), polymyxin B (300 units), trimethoprim (5 μg), or others as indicated by your professor (codes are printed on the top and bottom of the disc to tell you which antibiotic it contains–see **table 25.1**).
- Disc dispenser (if available)
- Tweezers
- Alcohol beakers for sterilizing tweezers
- Metric rulers (period 2)
- Bunsen burner and striker

## Procedure

### Period 1

1. Use a sterile swab **(figure 25.2)** to make a "lawn" of bacteria on each plate (completely cover the plate as shown in **figure 25.3**).
2. Discard your swabs in the biohazard container.

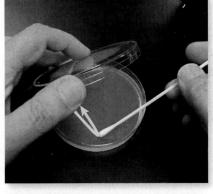

**Figure 25.2** Streaking the plate with a swab. ©Steven D. Obenauf

Use your swab to make parallel streaks in one direction that are close together.

Rotate the plate 90 degrees and make another set of parallel streaks.

Rotate the plate 45 degrees and make a final set of parallel streaks.

**Figure 25.3** Steps for using a swab to make a lawn of bacteria.

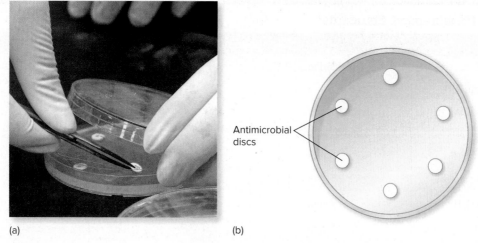

(a)                                                                      (b)

**Figure 25.4 Disc placement. (a)** Placing discs on a plate with tweezers; **(b)** final pattern of discs after placing them on the plate with tweezers or a disc dispenser. (a) ©Steven D. Obenauf

Antimicrobial discs

3. Use sterile tweezers to evenly disperse the six different antimicrobial discs around the plate **(figure 25.4b).** You may find it helpful to put a dot with your Sharpie on the bottom of each plate where the discs will be placed. Or if a disc dispenser is available, you will use that instead of tweezers. Place the dispenser over your plate (remove the lid) and press down on the plunger. Skip to step 8. Your instructor may direct you to use fewer than six discs.

4. Sterilize your tweezers by dipping them in alcohol and lighting the alcohol with the flame of your Bunsen burner. **Hold the tip of the tweezers below your hand** to avoid injury. Leave the tweezers in the burner flame only long enough to light the alcohol–do not hold them in the flame.

5. Pick up a cartridge (container) of discs.

6. Push against one edge of the disc at the opening of the cartridge, so that it sticks out slightly from the other side.

7. Grasp the disc with the tweezers and carefully place it on the surface of the plate **(figure 25.4a).**

8. Tap the disc down gently with the tweezers to make sure it is in good contact with the plate surface.

9. Sterilize the tweezers again and repeat until all three plates have the same pattern of antimicrobial discs on them (as shown in figure 25.4b).

10. Place the plates in the incubator lid side down.

## Period 2

1. Examine your plates for zones of inhibition, which will appear as a clear area of no growth around the disc visible against the lawn of bacteria covering the plate **(figure 25.5).** Measure the diameter of the zones in millimeters (mm). If you cannot measure the diameter directly (due to zones overlapping or intersecting the plate edge), then measure the radius and multiply by 2 to calculate the diameter.

2. Compare your zone of inhibition measurements to a standard **(figure 25.6** and **table 25.1)** to determine if the bacteria were sensitive or resistant to the antibiotic (or the result was intermediate).

3. Record your results.

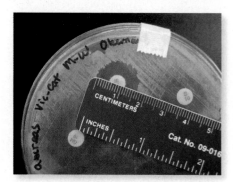

**Figure 25.5** Measuring zones of inhibition. ©Steven D. Obenauf

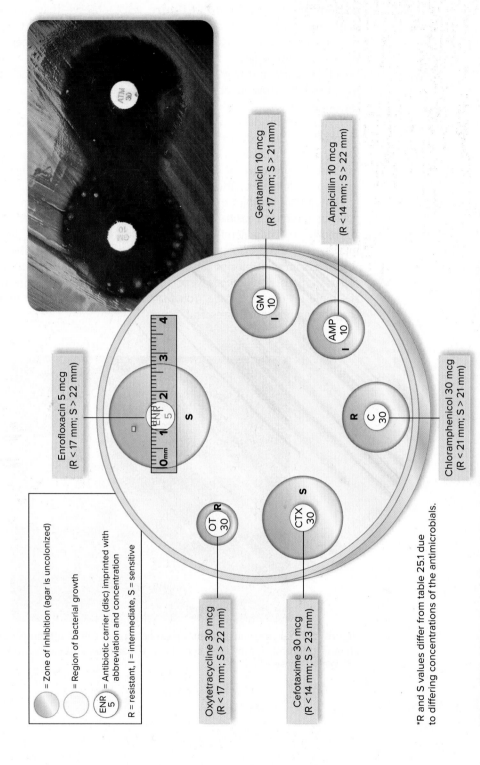

**Figure 25.6 Measuring and interpreting zones of inhibition.** Sensitive, intermediate, and resistant zones are illustrated. ©Kathleen Talaro

Gentamicin 10 mcg
(R < 17 mm; S > 21 mm)

Ampicillin 10 mcg
(R < 14 mm; S > 22 mm)

Enrofloxacin 5 mcg
(R < 17 mm; S > 22 mm)

Chloramphenicol 30 mcg
(R < 21 mm; S > 21 mm)

Oxytetracycline 30 mcg
(R < 17 mm; S > 22 mm)

Cefotaxime 30 mcg
(R < 14 mm; S > 23 mm)

= Zone of inhibition (agar is uncolonized)

= Region of bacterial growth

ENR 5 = Antibiotic carrier (disc) imprinted with abbreviation and concentration

R = resistant, I = intermediate, S = sensitive

*R and S values differ from table 25.1 due to differing concentrations of the antimicrobials.

## Results and Interpretation

Your three plates will probably not look the same. Different antimicrobial drugs have different mechanisms of action and so will not work uniformly with all bacteria. The drugs we are testing in this exercise have a number of targets, including the cell wall, the cell membrane, protein synthesis, and nucleic acid synthesis. In **figure 25.7,** observe that vancomycin and penicillin, antibiotics that are the drugs of choice for certain infections, have no zone and a very small zone, respectively. Differences often occur between gram-negative and gram-positive bacteria. Some antibiotics are broad-spectrum and may, in fact, work well with all the bacteria we test. In contrast, narrow-spectrum antimicrobials only work against one organism or group of organisms.

Measure the diameter of the zones in **millimeters,** not centimeters. If you can't get a diameter (this sometimes happens with large zones or zones that run together), measure the radius starting from the center of the disc to the edge of the zone of inhibition. Multiply the radius by 2 to get the diameter. Obviously, a disc with no zone of inhibition around it contains an antibiotic that particular bacteria are resistant to. A large zone of inhibition probably means that the bacteria are sensitive to that antibiotic. To *truly* determine if the antibiotic has worked, however, you must compare the diameter you measured to the standard table. An 18 mm zone for some antibiotics means that the bacteria are sensitive, while for others it does not. Record your measurements and the interpretation in the table on the next page. You may also wish to use your mobile device to photograph your plates and zones.

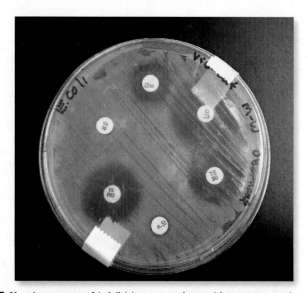

**Figure 25.7** Varying zones of inhibition on a plate with streptomycin, chloramphenicol, tetracycline, penicillin, gentamicin, and vancomycin discs.
©Steven D. Obenauf

# Antimicrobial Susceptibility Testing (Kirby-Bauer Method)

## Your Results and Observations

Record your results in the table. If instructed, you may photograph your plates with your mobile device.

| | | Patient B | | Patient S | | Patient X | |
| | | | Result | | Result | | Result |
| Antibiotic | Code | Zone (mm) | S/I/R* | Zone (mm) | S/I/R | Zone (mm) | S/I/R |
|---|---|---|---|---|---|---|---|
| Azithromycin | AZM-15 | | | | | | |
| Ciprofloxacin | CIP-5 | | | | | | |
| Penicillin | P-10 | | | | | | |
| Polymyxin B | PB-300 | | | | | | |
| Tetracycline | TE-30 | | | | | | |
| Trimethoprim | TMP-5 | | | | | | |
| | | | | | | | |
| | | | | | | | |

*S = sensitive, I = intermediate, R = resistant.

## Interpretation and Questions

1. Based on your results, which patient's infection appears to be the most difficult to treat? Why?

_____

_____

_____

_____

**2.** With which antibiotic should Patient S be treated?

_____

_____

_____

_____

**3.** Which two antibiotics appear to have the most similar spectra of activity (be effective or not on the same organisms)?

_____

_____

_____

_____

**4.** With which antibiotic should Patient X be treated?

_____

_____

_____

_____

**5.** In preparing your plate for testing, you spread the inoculum first in one direction, then in another, and finally in a third direction. Why was it important to spread the inoculum in this manner?

_____

_____

_____

_____

# ELISA and Using Nucleotide BLAST (Dry Lab)

## CASE FILE

A 14-year-old boy was brought to the office of his primary care physician suffering from the onset of sudden, involuntary muscle jerks. These twitches had begun the week before and were ongoing and gaining in severity. The patient had been adopted from the Philippines as a 2-year-old toddler. His early medical history, as indicated in his adoption records, had been unremarkable. Post-adoption, he had received all routine child-

Source: Center for Disease Control (CDC) - PHIL

hood immunizations. He had been apparently developmentally normal and healthy until the year before, when the patient started struggling in school. Teachers noted the patient was drifting off to sleep in class, had difficulty with homework, and was socially withdrawing from classmates. In response to the teacher's concerns, the patient's parents had sought counseling for their son. Counseling was nonproductive, however, and the patient was referred to his internist for medical evaluation.

A general medical examination conducted by the patient's internist returned normal results for all standard tests. The patient was then referred to a neurologist for further examination. Serum and blood chemistries were normal. Urine toxicology levels were negative. HIV, cytomegalovirus, and Epstein-virus serology tests were negative. Tests for hereditary and meta-bolic disorders were negative. Levels of the heavy metals arsenic, mercury, and lead were normal. An MRI detected bilateral periventricular white matter lesions. An electroencephalogram (EEG) showed a diffuse background with sharp spikes and exhibited high-voltage bursts occurring every 10 seconds, which appeared along with the myoclonic jerks. Testing of cerebrospinal fluid removed by a lumbar puncture revealed normal glucose levels (65 mg/dl), normal protein levels (47 mg/dl), and no bacterial cells. Because of the characteristic EEG, the patient's blood serum was tested for measles virus (MV) IgG. If the level of MV-IgG in the blood serum is elevated, a tentative diagnosis of subacute sclerosing panencephalitis can be made. Samples of the CSF were sent to the Centers for Disease Control and Prevention (CDC) for confirmation of the diagnosis.

The patient was discharged from the hospital. Over the following months, his motor and cognitive skills continued to decline. He became unable to sit and eventually could not walk. His speech became infrequent, random, and meaningless, and his behavior became more aggressive. He started to sleep more and experienced bouts of tachycardia and hyperthermia. He was admitted to a nursing home to ensure 24-hour care. After 8 months in the nursing home, he passed away.

In subsequent interviews with the patient's family, it was learned that he had never expressed overt signs of measles. His early medical history, before his adoption, did not indicate exposure to or infection with measles. The patient had, during the past 10 years, traveled out of the country (China, Philippines, and Japan) to several locations where vaccination for MV is still not widespread. The measles virus is still endemic in these countries, and viral outbreaks still commonly occur. Additionally, the patient had visited a large theme park in his home state, during which time one of the few recent measles outbreaks in the United States was documented to have occurred. This case of purported subacute sclerosing panencephalitis (SSPE) is being attributed to a subclinical case of measles, with the initial infection linked to the patient's exposure to the virus prior to adoption, during foreign travel in the last 10 years, or during the well-documented MV outbreak in the United States.

---

### LEARNING OUTCOMES

At the completion of this exercise, students should be able to

- describe the use of ELISA to identify specific antibodies in body fluids.
- align nucleotide sequences.
- demonstrate the use of BLAST or a similar identification program to identify a virus.
- describe the signs of measles virus infection.

---

## Background

Measles is a highly contagious, acute respiratory infection caused by the measles virus, a member of a Paramyxovirus family, genus Morbillivirus. It is a small, single-stranded RNA virus. The respiratory epithelium is the primary site of entry. Before the widespread use of the measles-mumps-rubella (MMR) vaccine (1963), the CDC estimates, in the United States between 3 and 4 million people contracted measles every year. Of those infected each year, 400-500 people died, many more developed encephalitis, and more than 40,000 patients were hospitalized. The MMR vaccine–administered in two doses, the first dose between 12 and 15 months and the second between 4 and 6 years of age—is 97% effective in preventing measles. In less than 70 years, the MMR vaccination program has decreased the incidence of measles cases by more than 99%. In the United States today, most measles cases are associated with sporadic outbreaks linked with importation events involving unvaccinated individuals who contracted the virus abroad and then brought it into the country. The CDC reports that, between 2010 and 2016, the largest outbreak reported in the United States occurred in 2014, resulting in a total of 667 cases. The originally infected individual and the majority of people who were exposed to and contracted the virus had never been vaccinated. Measles remains a significant health concern in the world. The World Health Organization (WHO) estimates that measles accounts for 15 deaths every hour. Vaccination for measles has been a WHO imperative, and in 2015 the organization estimated that 85% of all children worldwide received one dose of the vaccine. By WHO estimates, the vaccination program prevented 20.3 million deaths in a 5-year period.

The nervous system is not the primary target of the measles virus. It is not considered a neurotrophic virus, although neurological complications such as encephalitis can result from acute measles infections. SSPE is a rare, but always fatal, neurological complication resulting from a long-term, persistent infection of a mutated measles virus. SSPE results from the reactivation and infection of the brain by the virus. In the United States, because of its active and effective vaccination program, the decline in active measles cases has led to a concomitant decline in SSPE cases. Fewer than a dozen cases of SSPE are recorded in the United States each year. Most of the cases occur in unvaccinated individuals and usually in young people. SSPE is also more common in males than females.

One of the methods used to identify MV antibodies is the enzyme-linked immunosorbent assay (ELISA). ELISA is a technique that exploits the specificity of antibodies and capitalizes on the catalysis activity of enzymes to quantitate molecules of interest in a sample. The value of ELISA is that the assays are relatively easy to do, cheap, and fast.

Before the ELISA is discussed in more detail, a brief review of some terms and concepts is warranted. An **antigen** is a molecule, typically a protein (including other antibodies) or metabolite that elicits an immune response. An **antibody** is a protein produced in response to an antigen. The antibody will bind to a specific sequence found in an antigen. This specific sequence, or region, is called the epitope. Individual antibodies are very specific and will bind only to the epitope of their specific antigen.

**Enzymes** are proteins that catalyze reactions. Enzymes act on a substrate and produce a product. Enzymes are biological machines that, under constant conditions, work at a constant rate. The amount of product they produce, therefore, is proportional to the amount of enzyme present. The more enzyme present, the more product (which can be measured) that is produced. In most ELISAs, the enzymes that are used catalyze a reaction in which the substrate is a colorless compound and the product is colored (blue or yellow) **(figure 26.1)**. In the ELISA, the enzyme is bound (conjugated) to the antibody, which allows us to determine how much antigen is present. The antibody binds the antigen, and the enzyme bound to the antibody produces the product, which is then measured. The more antibodies that bind to antigens, the more enzymes are present, catalyzing reactions and producing products. In some cases, instead of an enzyme, the antibody is conjugated to a labeled compound of some type. The labeled compound can be a radioactive molecule or a colored or fluorescent compound that can be easily detected.

There are four basic ELISA design models: direct, indirect, sandwich, and competition or competitive. The model that is used depends on the research problem that needs to be addressed.

In a **direct** ELISA, the wells of the microtiter plate are coated with the antigen. The wells are then washed (a solution containing the antibody is added to the well) with the antibody-enzyme conjugate solution. The enzyme's substrate is then added, and the reaction is allowed to proceed. The microtiter plate is then placed on a plate reader, where the level of color produced (product of the enzyme reaction) in each well is measured.

**Figure 26.1** An ELISA showing a color reaction in the first three wells of the dilution series. ©Steven D. Obenauf

In an **indirect** ELISA, the wells are first coated with the antigen. The wells are then washed with an antibody specific for that antigen. This antibody is an unconjugated antibody, which is an antibody that lacks an attached enzyme. In order to identify and quantitate the antibody-antigen complex in the well, a second antibody is used. The second antibody, directed against the species that produced the first antibody and that has a conjugated indicator enzyme, is added next. For example, the well is coated with protein A (antigen). The well is washed with a rabbit anti-A antibody. To detect this antibody, a second antibody, an anti-rabbit antibody, is then added to the well. This anti-rabbit antibody has a conjugated enzyme. This is called an indirect test, because what are actually being detected are the antibodies, not the antigens directly.

In a **sandwich** ELISA, the antigen is sandwiched, or captured, between two antibody pairs. The well is first coated with an antibody called the capture antibody. The solution containing the antigen is then added to the well. The wells are rinsed and then washed with a second detection antibody. The detection antibody is conjugated to an enzyme or a label.

In the **competition,** or **competitive,** technique, antigen-specific antibodies are used to coat the wells. Then the solution containing the antigen is mixed with an antigen reagent that has been conjugated to an enzyme. This mixture of conjugated and unconjugated antigens is added to the well. The antigens bind to the antibodies and compete for binding sites. The wells are washed to remove unbound antigens. The enzyme substrate is added to the wells, and the test is conducted as with a normal direct ELISA. However, the interpretation of the results is quite different. High color production indicates low antigen content in the sample; if few antigen molecules are present in the sample, then more conjugated antigens bind to the well and produce more colored product. Conversely, if the sample has a high concentration of antigen, those molecules bind to the antibody sites. Fewer of the conjugated antigens bind, leading to less color development.

In this laboratory exercise, you will be doing an indirect ELISA on a cerebrospinal fluid sample taken from the patient. The presence of MV antibodies in the CSF indicates a central nervous system infection. If the antibody titer is sufficiently high and other clinical tests align, a diagnosis of SSPE can be made. The normal titer level in the CSF for MV-IgG is 5. For a diagnosis of SSPE, MV titer in the CSF is typically 10 to 100 times greater than normal.

Antibody titer is a measure of the level of antibodies to a specific antigen found in a body fluid. Practically, it is the greatest dilution of sample at which agglutination (interaction and precipitation of the antibody and antigen) occurs. In the clinical lab, a patient's sample is diluted using a set dilution sequence called a serial dilution. In this example, a given volume of sample is diluted in half and mixed to produce the first titer dilution in the dilution sequence. A sample is then taken from this tube and diluted in half again to create the next solution. This is repeated over and over again to produce the dilution sequence. The titer numbers reflect the dilution of the original sample; a 1:2 dilution indicates that the original sample was diluted in half by an equal volume, producing the 1:2 dilution. The titer is simply given as 2. The serial dilution would produce the following antibody dilutions: 1:2, 1:4, 1:8, 1:16, 1:32, 1:64, 1:128, 1:256, 1:512, and 1:1,024 dilutions. The possible titer values would be 2, 4, 8, 16, 32, 64, 128, 256, 512, and 1,024. Remember, the titer is defined as the highest dilution at which agglutination (the antigen and antibody bind and agglutinate) occurs. If agglutination is observed in the 1:16 well, but not in the 1:32 well, then the titer is 16. The higher the titer number, the more the sample has been diluted. The higher the dilution at which an agglutination reaction occurs, the greater the amount of that specific antibody in that body fluid.

 **Tips for Success**

- Add solutions to the wells carefully. Do not touch the sides of the well or scrape the bottom of the well with the pipette tip.
- Allow the solutions to incubate in the wells for the full time indicated in the procedure.
- Rinse the wells carefully; you do not need to eject the wash solutions in a "jetstream."
- Change the pipette tips when instructed to do so, and do not handle the pipette tips or other materials with your bare hands.

## Part I

### Materials

- Disposable microtiter plate
- Antigen (AG)–microfuge tube containing measles antigens
- Positive control (C++)– microfuge tube containing IgG-MV
- Negative control (C)
- Patient sample titrations (1:2, 1:4, 1:8, 1:16, 1:32, 1:64, 1:128)

- Secondary conjugated antibody
- Paper towels
- Enzyme substrate
- Micropipette (20-200 µl)
- Micropipette tips (1-100 µl)
- Transfer pipette
- Beaker
- Wash buffer
- Timer (optional)
- Marker

### Procedure

1. Label your microtiter plate as shown in **figure 26.2.** Your plate may differ, so label your plate as indicated by your instructor. The first two wells are the positive controls. The next two wells are the negative controls. The fifth well is empty. The remaining seven wells are dilutions of the patient's serum.

2. Use the micropipette to transfer 50 µL of antigen to the positive control and negative control wells in row A and into the patient sample wells in rows A and B. Discard the pipette tip.

3. Set the timer for 5 minutes. Wait 5 minutes. During this time, the antigen is binding to the plastic sides and bottom of the wells.

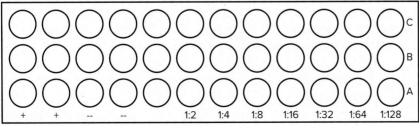

**Figure 26.2** Diagram of a typical microtiter plate.

4. After this incubation period, invert the microtiter plate over a stack of paper towels. Tap the microtiter plate gently on the paper towels. The paper towels will help wick away the fluid in the wells. Flip the microtiter plate back upright. Discard the wet paper towels.

5. Carefully add the wash buffer to each well using the transfer pipette. Do not overfill the wells.

6. Invert the microtiter plate onto a stack of paper towels again and tap gently. Flip the plate over again. Discard the wet paper towels. Repeat steps 5 and 6.

7. Place a new pipette tip onto the micropipette, and then transfer 50 μL from the positive control microfuge tube to the well labeled "+." Withdraw another 50 μL from the microfuge tube labeled "positive control" and transfer that to the second well labeled "+." Discard the pipette tip.

8. Place a new pipette tip onto the micropipette, and then transfer 50 μL from the negative control microfuge tube to the well labeled "–." Withdraw another 50 μL from the microfuge tube labeled "negative control" and transfer that to the second well labeled "–." Discard the pipette tip.

9. Place a new pipette tip onto the micropipette, and then transfer 50 μL from the 1:2 microfuge tube to the well labeled "1:2" in row A. Withdraw another 50 μL from the 1:2 microfuge tube and transfer that volume to the well labeled "1:2" in row B. Discard the pipette tip.

10. Place a new pipette tip onto the pipette, and then transfer 50 μL from the 1:4 microfuge tube to the well labeled "1:4" in row A. Withdraw another 50 μL from the 1:4 microfuge tube and transfer that volume to the well labeled "1:4" in row B. Discard the pipette tip.

11. Continue this process until all of the patient samples have been added to the wells. Start the timer (set for 5 minutes) once the last patient dilution has been added to the microtiter plate.

12. At the end of 5 minutes, invert the plate onto a stack of paper towels. Discard the paper towels.

13. Carefully add the wash buffer to each well using the transfer pipette. **Do not overfill the wells.**

14. Invert the microtiter plate onto a stack of paper towels again and tap gently. Flip the plate over again. Discard the wet paper towels. Repeat this wash cycle one more time (steps 13 and 14).

15. Using a fresh pipette tip, transfer 50 μL of the secondary conjugated antibody to each well using the micropipette.

16. Set the timer for 5 minutes to allow the antibody to bind to the MV-IgG.

17. Repeat the wash cycle (steps 4 and 5) three times.

18. After the final wash cycle, use the micropipette, with a new pipette tip, to add 50 μL of the enzyme substrate to the microtiter plate wells.

19. After 5 minutes, record the results. Your instructor will provide further information on how to read your results.

# ELISA and Using Nucleotide BLAST (Dry Lab)

## Results

Draw your results on the following figure.

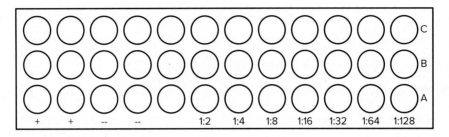

1. What is the titer of the patient's serum?

_____
_____
_____
_____

2. Can a diagnosis of SSPE be made based on this titer?

_____
_____
_____
_____

3. What are the advantages of using an ELISA as a clinical test?

_____
_____
_____
_____

**4.** For the ELISA used in class, name the conjugated enzyme and product and describe how the color reaction occurred.

_____

_____

_____

_____

**5.** What is the function of the enzyme used in the ELISA?

_____

_____

_____

_____

**6.** Explore reputable clinical resources and answer the following:

    **a.** What is the definition of _incubation period?_ How long is the incubation period for measles?

_____

_____

_____

    **b.** What is the prodromal period? How long is the prodromal period?

_____

_____

_____

    **c.** What are the signs indicating the prodromal period?

_____

_____

_____

    **d.** How long after exposure does the measles rash appear?

_____

_____

_____

    **e.** Describe the appearance of the rash and it origination.

_____

_____

_____

    **f.** What are typical complications of measles?

_____

_____

_____

## CASE FILE (CONT.)

Prior to the patient's final hospitalization, his parents met with representatives from the state's local health care agency. Health care officials were interested in the etiology and origin of the initial measles infection. The patient had never exhibited, to the parents' knowledge, any overt signs of measles, leading state health officials to suspect that the patient had been infected as an infant prior to adoption and coming to the United States, or possibly the patient had been exposed more recently and experienced a subclinical infection linked to one of the recent U.S. outbreaks. Given these two options, one possible avenue to explore these alternatives would be to identify the virus strain that caused the infection. Wild-type (a variety typically found in nature) measles virus strains vary from country to country and occurrence to occurrence. The National Institutes of Health through the National Center for Biotechnology Information (NCBI) maintains the genome bank for many organisms and viruses. If the virus causing the patient's infection could be isolated and its genome sequenced, it would be possible to identify its virus strain/genotype and then, using epidemiological techniques, track the infection to its source. Presented with this option, the parents agreed to allow CDC representatives to remove appropriate samples for virus isolation during autopsy.

Source: Center for Disease Control (CDC) - PHIL

At autopsy, a small amount of white matter was removed from the brain for analysis. The material was sent to the CDC. Measles virus RNA was isolated and purified. The material was sequenced and reverse transcribed.

## Background

Wild-type measles viruses are divided into 19 groups, or genotypes. These genotypes are based the nucleotide sequences of the hemagglutinin (H) and nucleocapsid protein (N) genes. Both of these genes are highly variable regions of the genome. According to the CDC, the nucleocapsid protein gene includes 450 nucleotides, coding for 150 amino acids. This region exhibits significant variation between genotypes. The reference strain for each genotype is the first strain identified with the specific unique genome. The total viral genome includes six genes: nucleocapsid protein (*N*), hemagglutinin (*H*), fusion (*F*), matrix (*M*), large protein (*L*), and phosphor (*P*).

Although MV-nucleic acid isolated from the brain tissue of patients with SSPE reveals mutations in several of these genes (*M, N, H, F*), more mutations seem to be associated with the *M* gene. The matrix, or M, protein plays a key role in the assembly of the virus. Researchers have suggested that the body's immune system might not respond to the virus because of these mutations, leading to the development of SSPE.

## Activity 2

The following lines are some of the bases found in the hemagglutinin gene of the virus isolated from the patient. These are fragments of variable lengths. The entire segment of the genome given here is 88 nucleotides. Your first task is to align the fragments correctly. Nucleic acids are directional molecules; for your purposes, the molecule will start at the 5′ (5 prime) end. To align the fragments, photocopy or scan this page, cut out each line of nucleotides, and then match up the identical sequence of bases. Your final sequence should be 88 nucleotides long.

> 5′ ctgagtctgacagttga
> 5′ catacggggtcttgtctgttaa
> 5′ gttaatctga
> 5′ agttgagcttacaatcaagat
> 5′ aggattccttcatacg
> 5′ agattgcttcaggattcgggccattga
> 5′ gtctga
> 5′ atcaagattgcttcaggatt

Now that you have aligned your sequence, write your sequence here:

_____

_____

_____

_____

Next, you will use this sequence to identify this virus using the BLAST tool at the National Center for Biotechnology Information (NCBI). BLAST is an abbreviation for "basic local alignment search tool." This free tool will search for and align any input sequence of nucleotides or amino acids with any available sequences in the published databases.

1. The NCBI website address is https://blast.ncbi.nlm.nih.gov/Blast.cgi. Once at that site, click on the box labeled "Nucleotide BLAST."
2. The next screen has five webpage tabs. The default tab is the blastn tab. Blastn is the page for blasting nucleotide (n) sequences.
3. Copy your 88 nucleotide sequence into the top box in "Enter accession number (s), gi(s) or FASTA sequence(s)."
4. In "Choose Search Set," the default "Others" should be selected. That is the correct selection.
5. Click on the "Blast" button at the lower left corner of the page.
6. The screen will immediately change to let you know the search has begun. The screen will update every few seconds. Your results should return within a minute, depending on your Internet connection speed. If you do not get a result, you need to reexamine your nucleotide alignment or retype your nucleotides into the submission box.

**7.** At the bottom of the page, you will see the top 100 virus genotypes and strains that produce the closest alignment.

Which measles genotype and strain gave the closest alignment?

_____

_____

_____

_____

Look to the far right of the table. What is the percent identification value returned for this sequence?

_____

_____

_____

_____

| Country | | Country | |
|---------|-----|---------|-----|
| Argentina | ARG | Liberia | LBR |
| Armenia | ARM | Libya | LBY |
| Bangladesh | BGD | Mexico | MEX |
| Belgium | BEL | Paraguay | PRY |
| Brazil | BRA | Peru | PER |
| Bulgaria | BGR | Philippines | PHL |
| Cambodia | KHM | Poland | POL |
| Cape Verde | CPV | Portugal | PRT |
| Chile | CHL | Puerto Rico | PRI |
| China | CHN | Singapore | SGP |
| Colombia | COL | Spain | ESP |
| Denmark | DNK | Thailand | THA |
| Ecuador | ECU | Turkey | TUR |
| El Salvador | SLV | Uganda | UGA |
| Ghana | GHA | United States | USA |
| Japan | JPN | Venezuela | VEN |
| Lebanon | LBN | Viet Nam | VNM |
| | | Zimbabwe | ZWE |

Measles virus isolates are named using a convention that includes the sample's country of origin. The virus name includes a three-letter abbreviation for the country of origin. A sample of some of the country abbreviations is given here. Using this table, and the identity of the measles virus hemagglutinin gene, from which country did this virus originate?

_____

_____

_____

_____

Knowing what you know now about this virus and this patient's history, what is your explanation for when and how this patient contracted this virus?

_____

_____

_____

_____

# Analysis of DNA: Electrophoresis

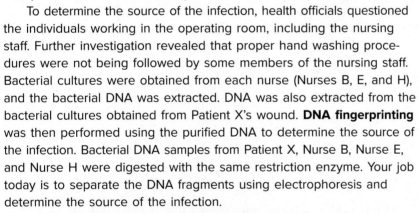

## CASE FILE

In late August, a 37-year-old woman (Patient X) developed a massive infection of a surgical wound after a routine surgical procedure. A Gram stain of the wound culture revealed the presence of a gram-negative rod. Another man at the same hospital was also infected with a gram-negative bacterium. The two patients confirmed to be infected were both at the same surgical unit on the same day. Five other patients, who also had surgery on the same day as the previous patients, were tested; three were found to be infected. To determine the genus and species of the gram-negative bacteria, multitest systems (EnteroPluri-Tests) were inoculated with samples from all infected patients.

To determine the source of the infection, health officials questioned the individuals working in the operating room, including the nursing staff. Further investigation revealed that proper hand washing proce-dures were not being followed by some members of the nursing staff. Bacterial cultures were obtained from each nurse (Nurses B, E, and H), and the bacterial DNA was extracted. DNA was also extracted from the bacterial cultures obtained from Patient X's wound. **DNA fingerprinting** was then performed using the purified DNA to determine the source of the infection. Bacterial DNA samples from Patient X, Nurse B, Nurse E, and Nurse H were digested with the same restriction enzyme. Your job today is to separate the DNA fragments using electrophoresis and determine the source of the infection.

## LEARNING OUTCOMES

At the completion of this exercise, students should be able to

- describe the mechanism of electrophoresis as it is applied to DNA.
- perform DNA fingerprinting.
- interpret the results of a stained gel following a DNA finger-printing procedure.

## Background

Restriction endonucleases are enzymes that cut DNA at a specific sequence of nucleotides known as a restriction site. There are many different endo-nucleases, each with a different restriction site. DNA samples with different

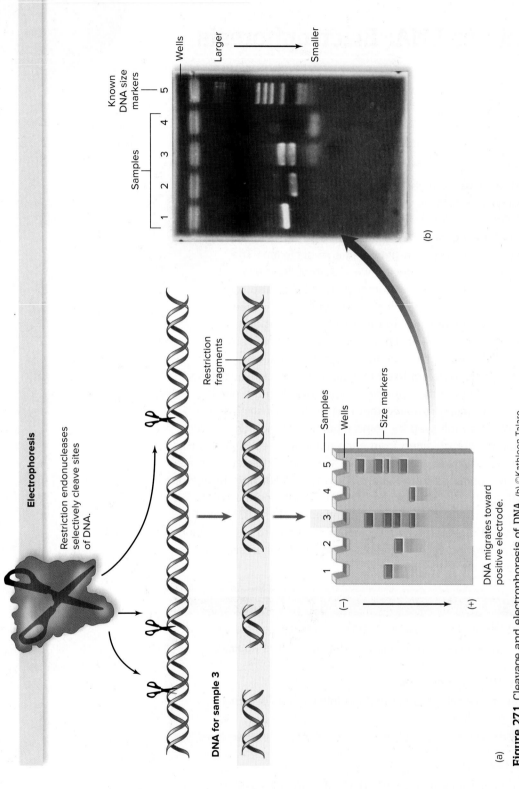

**Figure 27.1** Cleavage and electrophoresis of DNA. (b) ©Kathleen Talaro

nucleotide sequences, such as samples from two different strains of the same bacteria, will be cut differently by the same enzyme and yield different amounts and sizes (restriction fragment polymorphisms [RFLPs]) of fragments. One way to examine these fragments is by electrophoresis. To perform **electrophoresis,** digested DNA samples are mixed with loading dye and loaded into the wells of an 0.8% agarose gel. An electric field applied across the gel causes the negatively charged DNA fragments to move from their origin (the sample well) through the gel toward the positive electrode. The gel matrix acts as a sieve through which smaller DNA molecules move faster than larger ones; fragments of differing sizes separate into bands during electrophoresis **(figure 27.1).** The pattern of bands (the "fingerprint") produced for each sample is made visible by staining with a dye (in this exercise, SYBR Green) that binds to the DNA molecule. DNA electrophoresis, (also known as fingerprinting) is a useful tool for epidemiologists to determine the source of infection and identify the organism causing a particular infection.

## Materials (per Team of Two or Four)

- Bacterial DNA samples (Patient X; Nurses B, E, and H)
- Melted 0.8% agarose
- 1X Tris/borate/EDTA (TBE) buffer
- Gel tray, base, comb, and casting gates
- Gel tray lid and cables
- 1 µL, 4 µL, and 10 µL micropipettors and tips
- Tube racks
- Microcentrifuge

- Gloves
- Loading dye (Promega 6X Blue/Orange Loading Dye)
- SYBR Green DNA stain (Lonza) 1:1,000
- Power supply
- Funnel
- Ziploc bags
- UV light box (period 2)
- Gel Doc unit (gel documentation system) if available (period 2)

## ✔ Tips for Success

- Wear gloves when handling the microcentrifuge tubes. Nucleases on your hands can degrade the DNA in your samples.
- If the gels were not poured before class, pour the agarose gel as soon as you get to your benches. Agarose takes at least 10 minutes to solidify.
- As you are dispensing the samples into the wells, use your index finger of your non-pipetting hand to steady the tip. If the tip punctures the well, the sample will be lost.
- When dispensing the samples, slowly and steadily press the thumb down on the plunger. If you dispense the sample too quickly, it will not stay in the well.
- Make sure to place your gel in the electrophoresis chamber with the wells toward the black electrode.

## Procedure

### Period 1

#### Prepare DNA samples

1. Transfer 4 µL of 6X loading dye into each DNA sample (X, B, E, and H) **(figure 27.2)** using a micropipette. Pipette the dye onto the **side** of the tube, not the bottom.

2. Add 1 µL of SYBR Green to each sample (your professor may do this for you). Close the tube tops and mix by tapping the tube

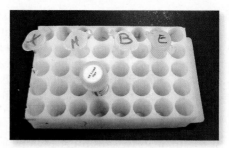

**Figure 27.2** Samples to be loaded into the gel. ©Steven Obenauf

bottoms on lab bench. Spin down briefly in a microcentrifuge (if available). Mix the contents after centrifuging by gently tapping the side of the tubes.

## Casting 0.8% agarose gel (to save time, this may be done for you before class starts)

**Figure 27.3** The gel unit and its components. ©Steven Obenauf

1. If the gel unit is not already set up, place the gel tray in the casting base. Place a gel casting gate on each end of the tray–make sure you have a good seal.
2. Insert the comb into the comb slots near one end of the gel.
3. The wells formed by the comb should be at the negative (black) electrode **(figure 27.3)**.
4. Pour warm agarose into the gel bed. Gel will become cloudy as it solidifies (about 10 minutes). DO NOT MOVE OR BUMP THE CASTING TRAY WHILE THE GEL IS SOLIDIFYING.
5. When agarose has solidified, unseal the ends of the gel tray (lift up the gel casting gates). Gently remove the comb. DO NOT THROW THE COMB AWAY–rinse it and return it to your workstation.

## Loading gel and electrophoresis

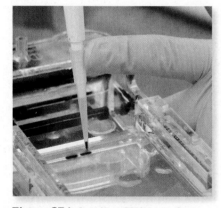

**Figure 27.4** Loading DNA samples into the wells. ©Steven Obenauf

1. Fill the gel box with TBE buffer by first filling the reservoirs at either end of the gel box and then by slowly pouring buffer until the buffer just covers the entire surface of the gel. Make sure there are no air bubbles in the sample wells.
2. Use a micropipettor to SLOWLY load 10 µL of each sample into a separate well in the gel (you may have some empty wells). Use two hands to hold the micropipettor. One hand should grasp the micro-pipettor near the top and one finger from the other hand should stabilize the bottom as shown in **figure 27.4.** Use a fresh tip for each tube. Place the pipette tip just inside the top of the sample well. The tip will be below the surface of the buffer. The high density of the loading dye will cause the sample to sink to the bottom of the well. DO NOT PUNCH THE TIP OF THE PIPETTE THROUGH THE BOTTOM OF THE SAMPLE WELL. **Load the samples using the pattern (B, E, H, space, X) shown in the lab report.**
3. Place the lid on the base and connect the electrical leads to a power supply, anode to anode (red–red) and cathode to cathode (black–black). Each power supply has output jacks for more than one gel unit, so teams can share a power supply if needed.
4. Turn power supply on and set the voltage to 90 to 110 volts using the up/down arrows on the front panel **(figure 27.5)**. Press the

**Figure 27.5** Power supply. ©Steven Obenauf

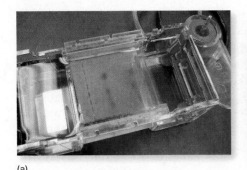
(a)

(b)

**Figure 27.6 Electrophoresis. (a)** Location of colors from the loading dye after electrophoresis is completed. **(b)** Removing the gel tray after electrophoresis is completed. (a-b) ©Steven Obenauf

Run/Pause button of the power supply to start running electricity through your gel. Run for about 40 minutes. The blue color of the loading dye should move out of the sample wells and across the gel, so you can track its progress; the farthest color shows the position of the smallest (fastest) molecules. Good separation will have occurred when the bromophenol blue dye (in the loading dye) front has moved 4 to 8 centimeters (cm) from the sample wells and the yellow band has gone all the way across the gel **(figure 27.6a)**. The dye may separate into blue and yellow bands.

5. Turn off the power supply, disconnect the leads from the power supply, and remove the lid.

6. Carefully remove the gel tray from the base **(figure 27.6b)** and slide gel into the Ziploc bag if you plan to observe it during the next lab period. Label the bag with your group name and lab day and time. The gel will continue to stain and will be observed during the next class period. It **must remain in the refrigerator in the dark.** (Option 2: If there is enough time, place the gel on the UV light box or in the Gel Doc unit immediately after finishing the electrophoresis.)

## Period 2 or End of Period 1

### Observing the gel "fingerprint"

1. Observe your gel by placing it into the Gel Doc unit and recording the image with a computer **(figure 27.7a)** or by placing the gel on a UV light stand and observing the glowing bands **(figure 27.7b)**.

2. Examine the stained gel or a photograph of it and sketch the banding patterns.

## Results and Interpretation

When using DNA fingerprinting epidemiologically to trace infections, interpreting the gel involves comparing the pattern of the bands in the samples from infected people or samples to the pattern of bands in samples from the potential sources of the infection. The lane for Sample X is the infected patient. You will compare the bands in this lane to those in lanes B, E, and H—the potential sources of the infection **(figure 27.8).**

(a)

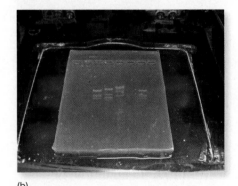
(b)

**Figure 27.7** Observing the results of the gel with **(a)** a computerized documentation system or **(b)** a UV light unit. (a-b) ©Steven Obenauf

Successful gels photographed as either positive or negative images are shown in **figure 27.8a, b.** You can ruin a good gel by handling it improperly. **Figure 27.8c** shows a gel where the surface was contaminated with material, such as glove powder, which glows under UV light. If the samples are loaded into the wells too quickly and flow out, your gel may look like **figure 27.8d.** If you position the gel improperly (reversing it) relative to the positive and negative electrodes, you will be rewarded with a gel that also looks like figure 27.8*d*. If the concentration of the sample or the voltage used is too high, the resulting gel may look like **figure 27.8e.**

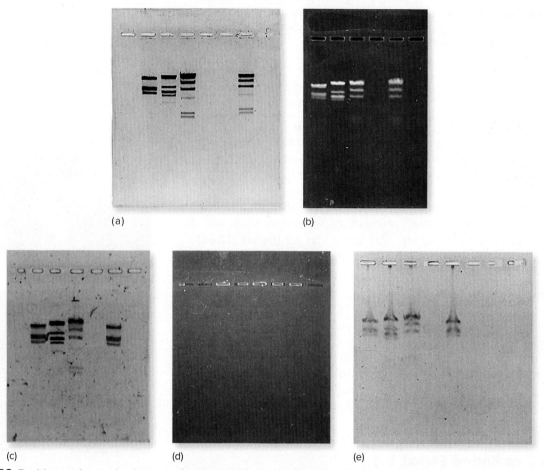

(a)  (b)

(c)  (d)  (e)

**Figure 27.8** Positive and negative images of stained DNA electrophoresis gels. (a-e) ©Steven Obenauf

# Analysis of DNA: Electrophoresis

## Your Results and Observations

Draw or photograph your gel with a mobile device.

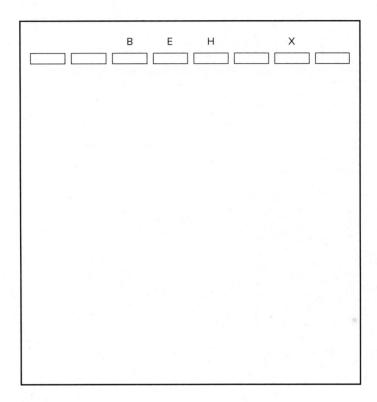

## Interpretation and Questions

**1.** What is the function of restriction endonucleases?

_____

_____

_____

_____

**2.** Explain the process of electrophoresis.

_____

_____

_____

_____

**3.** What is the purpose of DNA fingerprinting from an epidemiological standpoint?

_____

_____

_____

_____

**4.** Based on the preceding results, were any of the nurses responsible for the spread of this infection in the surgical wounds of Patient X? If so, which one? Explain your answer.

_____

_____

_____

_____

**5.** Your digest produces five distinct bands. Are all of the fragments within each band identical in composition?

_____

_____

_____

_____

EXERCISE

28

# Gene Transfer: Transformation

## CASE FILE

A 35-year-old man was diagnosed with type 2 diabetes and started insulin treatment in April 1982 with once-daily bovine and porcine insulin mixture. Eight days later, he began to see hives appearing shortly after he received his insulin dose. The severity of his hives increased and insulin was stopped 5 weeks later. In February 1983, treatment with purified porcine insulin was started. After his second injection, redness and swelling reappeared and persisted despite the patient being given oral prednisolone. In March 1983, he experienced generalized itching, vomiting, dizziness, and collapse due to shock 3 minutes after an insulin injection. Insulin was stopped again. Thirst, polyuria, and ketonuria from his diabetes returned, and twice-daily Humulin (recombinant human insulin) was started 2 days later as previous prick tests with human insulin showed no signs of an allergic reaction.

©MedicalRF.com RF

The use of bovine insulin for treating diabetes started in the 1920s, with a variety of improvements being made over the next few decades. Since their introduction, a small number of people have developed allergies to bovine insulin, porcine insulin, or both. The genetic engineering of *Escherichia coli* to produce human insulin gave an alternative source to those allergy sufferers.

### LEARNING OUTCOMES

At the completion of this exercise, students should be able to

- understand the importance of horizontal gene transfer in bacteria.
- describe the mechanism of transformation.
- be able to perform and interpret a transformation procedure in the laboratory.

## Background

The ability to pass on genetic material from generation to generation is essential for the survival of all species. Binary fission in bacteria generates offspring that are genetically identical to the parent cell (unless there is a mutation). To generate genetic diversity, bacteria utilize three forms of genetic transfer (sometimes called **horizontal gene transfer**).

**195**

The significance of this type of genetic transfer is that individuals in a population can easily spread or acquire new traits. The rise in the number of bacteria that are antibiotic resistant can be directly attributable to horizontal gene transfer. The three methods of genetic transfer in bacteria involve different mechanisms: In conjugation, bacteria are in direct contact using a sex pilus. DNA (plasmid or chromosomal) is transferred from the donor cell to the recipient through the pilus. Transduction is a process by which bacterial viruses called bacteriophages transfer DNA from one bacterium to another. In the process of **transformation,** bacterial cells take up DNA from the surrounding environment. The DNA is transported through the bacterial cytoplasmic membrane into the cytoplasm. Transformation requires cells to become **competent** (able to take in genetic material across their membrane).

In today's exercise, you will utilize cold $CaCl_2$ (calcium chloride) to make *E. coli* cells "competent." The calcium chloride will open pores in the cytoplasmic membrane of the *E. coli* cells, allowing much greater intake of a plasmid you will introduce. The plasmid you will introduce into *E. coli* is known as pUC18 and carries a gene (beta-lactamase) that encodes for resistance to the antibiotic **ampicillin.** Beta-lactamase inactivates ampicillin and other penicillin-based antibiotics. Normally, *E. coli* is susceptible to ampicillin, but if this plasmid is introduced, the *E. coli* will be able to grow in the presence of ampicillin. You will make two plates. The **"DNA+"** plate will contain *E. coli* cells that have been rendered competent with $CaCl_2$ and introduced to the plasmid. If the uptake of the plasmid is successful, the *E. coli* cells will be transformed from ampicillin susceptible to ampicillin resistant. The medium in the plate is nutrient agar with an ampicillin additive. The newly transformed *E. coli* cells will grow on the plates due to their acquired resistance. If one of the alternate procedures is used, the colonies may be a different color or glow green under UV light. The **"no DNA"** plate will contain *E. coli* that was rendered competent but *was not* introduced to extracellular DNA (pUC18). Due to the absence of the plasmid, there should be no acquired resistance to ampicillin, so the cells will not be able to grow on the plate.

Molecular biology exploits horizontal gene transfer to genetically engineer bacteria. For example, human proteins can be expressed using bacterial cells by introducing or inserting the gene for the desired human protein into the bacteria. The method often used to introduce DNA into bacterial cells is transformation. The DNA transferred is typically in the form of a plasmid. **Plasmids** are small, circular, self-replicating DNA molecules. They will continue to replicate once transferred into a new bacterium. These plasmids may also carry antibiotic-resistance genes. Those genes can be used as a selection tool for molecular biologists working with bacteria that have incorporated a plasmid.

## ✅ Tips for Success

- You will be using micropipettors set to three different volumes (100, 500, and 700 microliters [µL]). Check the settings as you use them to make sure you use the right one at the right step.

## Materials

- Competent *E. coli* cells (The instructor will make the cells competent by doing the following: Transfer 500 µL of $CaCl_2$ to the *E. coli*. Mix well. Using the same pipette, transfer the entire contents of the *E. coli*/$CaCl_2$ mix into the $CaCl_2$ vial, swirl to mix, and keep on ice. Cells will be ready for transfer in approximately 30 minutes and will be competent for up to 3 hours.)

- Ice bucket
- Heater block or water bath at 37°C
- 100-μL micropipettor and micropipettor tips
- 500-and 700-μL micropipettors and tips
- Microtube labeled "DNA+" (contains 5 μL of plasmid DNA)
- Microtube labeled "no DNA"
- Sharps container
- 2 plates of nutrient agar containing 75-μg/mL ampicillin (optional: may also contain X-gal and IPTG)
- Alcohol beakers
- Spreader bars
- Nutrient broth

## Procedure

### Period 1

1. Add 100 μL of competent cells to the "DNA+" tube.
2. Add 100 μL of competent cells to an empty sterile microtube labeled "no DNA."
3. Gently "tap" the tubes with your index finger to mix and place on ice **(figure 28.1a)** for 20 minutes.
4. After sitting on ice, remove the tubes and place them in the 37°C heat block **(figure 28.1b)** or water bath for 5 minutes. (This is known as heat shocking and will facilitate DNA uptake.)
5. Remove both tubes from the heat block and add 700 μL of nutrient broth to each tube.
6. Place both tubes BACK in the 37°C heat block and incubate for 20 minutes.
7. After incubation, remove 500 μL from each tube and put onto the appropriately labeled ampicillin plate ("DNA+," "no DNA"). Be sure to put the "DNA+" tube contents onto the plate labeled "DNA+" and the same for "no DNA." **Be sure to change the tips on the micropipettor between each transfer.** (It helps to do the "no DNA" transfer first.)
8. Sterilize the spreader bar; then use it to spread the cells across the plates **(figure 28.1c)**. **Be sure to sterilize the spreader bar well between plates.** Optional: If not using X-gal, divide one ampicillin plate in half and streak one side from the "DNA +" tube and one side from the "no DNA" tube.
9. Let the plates sit for a while, so the cells can adsorb onto the agar surface. Place the plates in the incubator, lid side down, and let them incubate until next lab period.

### Period 2

1. Observe your two plates ("DNA+" and "no DNA") and determine if the bacteria were transformed. Count the number of colonies (if X-gal was used, count blue and white colonies; if green fluorescent protein was used, count glowing colonies).

## Results and Interpretation

Look at the "DNA+" plate for colonies. The presence of colonies **(figure 28.2a, c)** indicates that the bacteria were resistant to the ampicil-

(a)

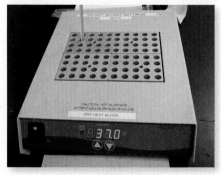
(b)

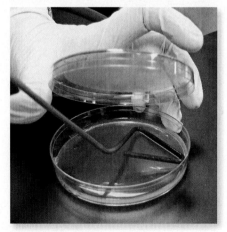
(c)

**Figure 28.1 Steps in the transformation process. (a)** DNA uptake on ice; **(b)** heat shocking; **(c)** spreading cells on the plate.
(a-c) ©Steven Obenauf

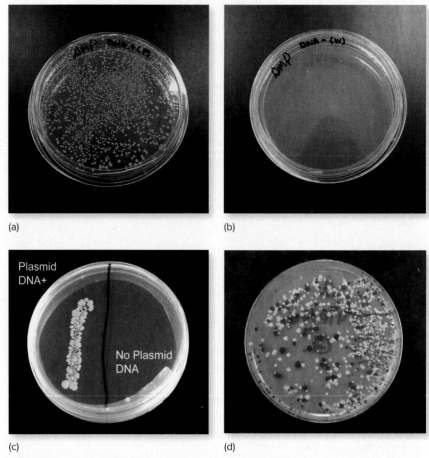

(a)

(b)

Plasmid
DNA+

No Plasmid
DNA

(c)

(d)

**Figure 28.2 Transformation plates.** Plates with ampicillin: **(a)** contains no X-gal or indicator; **(b)** prepared from a "no DNA" sample; **(c)** divided plate with DNA+ on one side and no DNA on the other; **(d)** has X-gal and indicator.
(a-d) ©Steven Obenauf

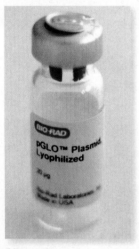

©Steven Obenauf

lin in the medium. The gene for ampicillin resistance was on the plasmid, giving the bacteria the ability to produce a beta-lactamase that destroys ampicillin. Absence of colonies could mean that the "DNA+" and "no DNA" tubes were switched or set up improperly.

Look at the "no DNA" plate for colonies. No growth on this plate **(figure 28.2b)** indicates that the organism could not destroy ampicillin to survive and grow on this ampicillin-containing medium. Colonies on the "no DNA" plate could mean that the medium did not contain sufficient amounts of ampicillin. The two sides of a divided plate **(figure 28.2c)** should be different, as plates (a) and (b) are.

If X-gal and plates with an indicator were used **(figure 28.2d)**, count blue (transformed by a plasmid containing an intact *LacZ* gene) and white colonies (transformed by a plasmid with an insert) on each plate. If green fluorescent protein (gfp) or pGLO was used, count the colonies that fluoresce under UV light (transformed) and do not glow or fluoresce (not transformed).

Name _____

Date _____

# Gene Transfer: Transformation

## Your Results and Observations

Draw or use a mobile device to photograph your results.

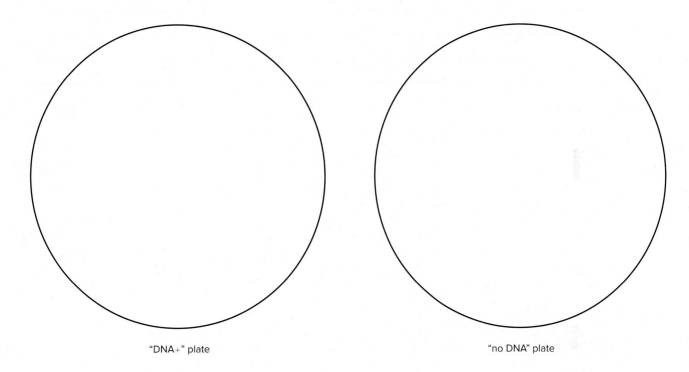

"DNA+" plate                                    "no DNA" plate

Count the number of colonies on the "DNA+" and "no DNA" plates.

| | "DNA+" plate | "No DNA" plate |
|---|---|---|
| Number of colonies | | |

## Interpretation and Questions

**1.** What are competent cells?

_____

_____

_____

**2.** Explain why the presence of antibiotics in the growth medium is important to the interpretation of the results.

_____

_____

_____

_____

**3.** The production of recombinant human insulin described in the Case File is a positive application of horizontal gene transfer. What negative situation (from the point of view of humans) has resulted from horizontal gene transfer?

_____

_____

_____

_____

# Parasitic Protozoa

## CASE FILE

Patient D, a woman who lives near the Appalachian Mountains, complained of having diarrhea. She reported to the nurse taking her history that she had been having abdominal pain that started a month ago. She also began having profuse, watery diarrhea at that time. The diarrhea was often explosive. Her dogs were exhibiting similar symptoms. Upon questioning, she stated that she had accidentally and the dogs had intentionally drunk water from the river below their home. Her workup included negative stool cultures for bacteria and a normal sigmoidoscope examination. A stool sample was sent to be examined for parasites, as you will be doing today. Under the microscope, many pear-shaped trophozoites with two bilateral nuclei, as well as oval cysts with two or four nuclei, were seen in her sample.

©MedicalRF.com RF

## LEARNING OUTCOMES

At the completion of this exercise, students should be able to

- understand the basic life cycle of parasitic protozoa and how protozoal diseases are transmitted.
- be able to recognize parasites from each of the four major categories of protozoa, the type of disease they cause, and the type of specimen in which they would be found.

## Background

Most protozoa are beneficial members of aquatic ecosystems, as described in exercise 5, "Protozoa." A small number are parasitic and some cause some very important diseases, leading to millions of infections and deaths each year. Many protozoa **(figure 29.1)** will spend part of their lives in the **cyst** stage (a tough, resistant, survival stage) and part in the **trophozoite** stage (the stage involved in feeding and reproduction). The cyst stage is often involved in disease transmission. The four major categories of protozoa are flagellates, amoebas, ciliates, and nonmotile protozoa.

A number of **flagellated** protozoa cause disease, including *Giardia*, *Trypanosoma*, and *Trichomonas*. **Giardiasis** is an intestinal infection transmitted when people or dogs drink the cyst of *Giardia* in contaminated water **(figure 29.2a, b)**. The cyst stage is resistant to chlorine, and this

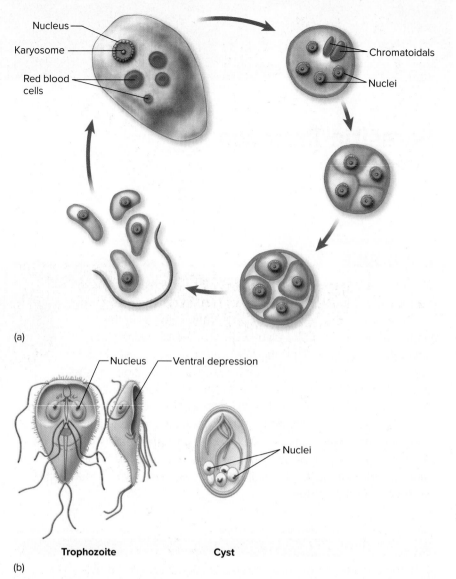

**Figure 29.1** Trophozoite and cyst stages in **(a)** *Entamoeba* and **(b)** *Giardia*.

disease is relatively common in the United States. It causes diarrhea, which can last a long time. Giardiasis is usually diagnosed by immunoassays or the examination of feces. **Sleeping sickness** (African trypanosomiasis) is found in central Africa and is transmitted by the bite of the tsetse fly. The parasite lives in the bloodstream for months or years, so a blood smear is used for diagnosis **(figure 29.2c).** When the *Trypanosoma* reaches the central nervous system, more serious symptoms arise, including coma and possibly death. In the Americas, a different species of *Trypanosoma* causes Chagas disease. **Trichomoniasis** is caused by the flagellate *Trichomonas* and is sometimes seen when the normal vaginal biota is altered. Diagnosis is usually made by looking microscopically for the organism in the discharge that it causes **(figure 29.2d).**

Amoebas that can cause disease include *Acanthamoeba*, which causes eye infections, and *Entamoeba*, which causes intestinal infections. **Amoebiasis** (amoebic dysentery) is an intestinal infection transmitted by ingesting the cyst of *Entamoeba* in food or water **(figure 29.3a).** The growth and colonization of the gut by the trophozoite **(figure 29.3b)** can cause severe diarrhea and is usually diagnosed by examination of feces.

The major **ciliated** protozoan associated with disease is *Balantidium* **(figure 29.3c, d). Balantidiasis** is an intestinal infection.

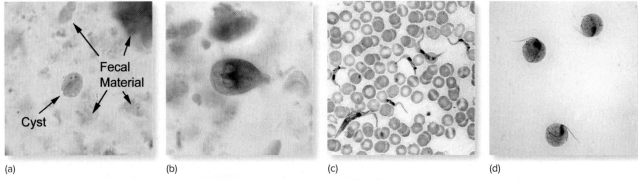

(a)    (b)    (c)    (d)

**Figure 29.2** Parasitic flagellates. **(a)** *Giardia* cyst in a fecal smear; **(b)** *Giardia* trophozoite in a fecal smear; **(c)** *Trypanosoma* in a blood smear; **(d)** *Trichomonas* in a vaginal smear. (a-d) ©Steven Obenauf

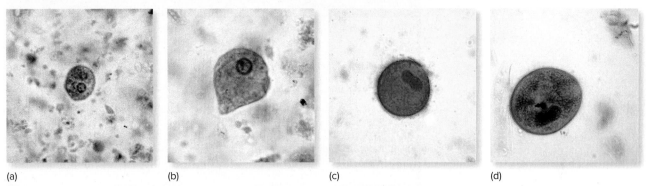

(a)    (b)    (c)    (d)

**Figure 29.3** *Entamoeba*, a parasitic amoeba, and *Balantidium*, a parasitic ciliate. **(a)** *Entamoeba* cyst in a fecal smear; **(b)** *Entamoeba* trophozoite; **(c)** *Balantidium* in a fecal smear; **(d)** *Balantidium* with a different stain. (a-d) ©Steven Obenauf

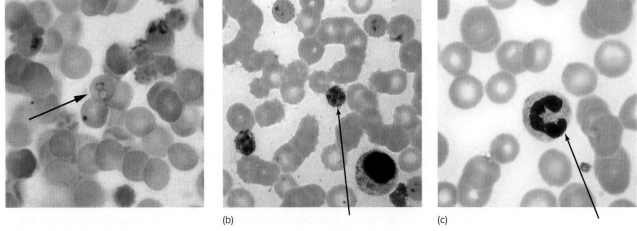

(b)    (c)

**Figure 29.4** *Plasmodium* and malaria. **(a)** Ring stage; **(b)** schizont stage (1,000x). **(c)** You may see this cell at the pointer in your blood smear as well. Is it a parasite? (a-c) ©Steven Obenauf

The apicomplexan or sporozoan protozoa are all **nonmotile** parasites and have complex life cycles. *Toxoplasma* and *Plasmodium* are two examples of this group. **Malaria** is caused by *Plasmodium* and found in many tropical areas of the world. It is transmitted by the *Anopheles* mosquito (which also serves as a host). It goes through a number of stages and locations once it gets inside the human body. Its major target is red blood cells (**figures 29.4** and **29.5**). Malaria is characterized by a fever that comes and goes in relation to the cycle of the parasite in the red blood cells.

Millions of deaths per year are caused by this disease. Many who die are children. Laboratory diagnosis of malaria is usually made by examining a blood smear, as you will do today.

## ✔ Tips for Success

- Use 1,000× magnification (oil immersion) for all of these organisms except *Balantidium*. For *Balantidium*, use 400×.
- Be careful not to get oil on your 40× (high dry) objective. Oil on the objective will make your image VERY blurry.
- Have the picture of the organism you are looking for visible next to you as you look under the microscope. Looking through a fecal (figure 29.2*a*) or blood smear when you don't know what the thing you are looking for *looks* like is not time well spent.

## Organisms (Prepared Slides)

| Organism | Disease | Sample |
|---|---|---|
| *Giardia lamblia* | Giardiasis | Fecal smear of trophozoites or cysts |
| *Trichomonas vaginalis* | Trichomoniasis | Vaginal smear |
| *Trypanosoma brucei* | Sleeping sickness | Blood smear |
| *Balantidium coli* | Balantidiasis | Fecal smear or intestinal section |
| *Entamoeba histolytica* | Amoebic dysentery | Fecal smear of trophozoites or cysts |
| *Plasmodium* spp. | Malaria | Blood smear |

## Procedure

1. In the *Giardia* fecal smears, you may see either the cyst or trophozoite stage. The cyst is oval in shape with two or four nuclei visible, while the trophozoite is somewhat pear-shaped with two nuclei. There are usually lots of *Giardia* present in the sample. The trick is to find one you can see well in between all of the other fecal material. Be patient.
2. For *Trichomonas*, you may see individual cells or clumps. The flagella usually show up well.
3. For *Trypanosoma*, you will see lots of red blood cells (RBCs) and a few white blood cells. There are usually a lot of clearly visible parasites as well. They will be between the RBCs.
4. For *Balantidium*, there are usually only a few on each slide. Find them at 40× or 100× first. They will be oval with a slightly bent nucleus. Observe them at 400×. Don't get fooled by other oval objects in your slide–they won't have a distinct nucleus as *Balantidium* does.
5. In the *Entamoeba* fecal smears, you may see either the cyst or trophozoite stage. It will probably take a while to find them with all the fecal material around. Cysts are round; trophozoites may be oval to irregular (amoeboid) in shape. Look for the circular nucleus with a very distinct edge. Go up and down with your fine focus.
6. For *Plasmodium*, you will need to look *inside* of the red blood cells for the parasite. This may take a while. Depending on the slide you have, you may see anything from the ring trophozoite stage to an infected RBC (schizont) about to rupture (figure 29.5).
7. Observe any other organisms your professor may assign.
8. Draw the organisms you observe.

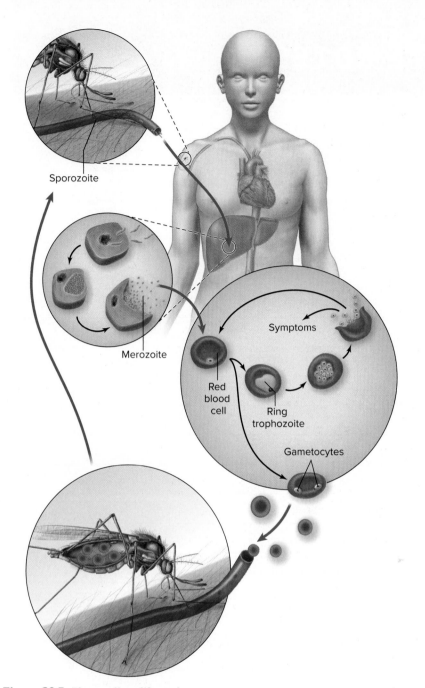

Sporozoite

Merozoite

Symptoms

Red
blood
cell

Ring
trophozoite

Gametocytes

**Figure 29.5** *Plasmodium* life cycle.

# NOTES

Name _____

Date _____

# Parasitic Protozoa

## Your Results and Observations

Draw or photograph with your mobile device the organisms you observed.

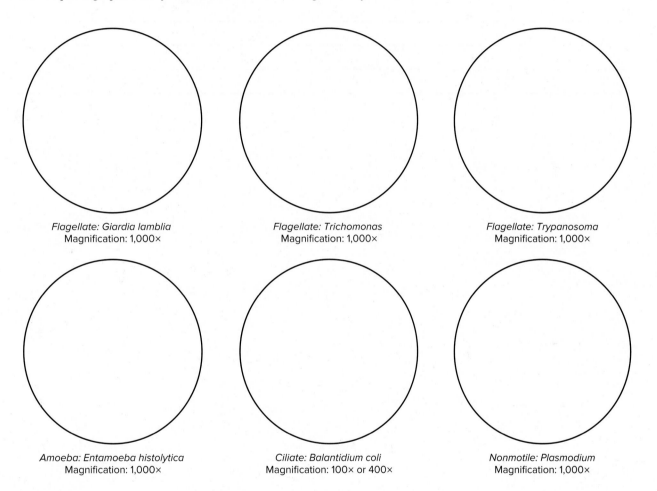

*Flagellate: Giardia lamblia*
Magnification: 1,000×

*Flagellate: Trichomonas*
Magnification: 1,000×

*Flagellate: Trypanosoma*
Magnification: 1,000×

*Amoeba: Entamoeba histolytica*
Magnification: 1,000×

*Ciliate: Balantidium coli*
Magnification: 100× or 400×

*Nonmotile: Plasmodium*
Magnification: 1,000×

## Interpretation and Questions

1. Based on the information in the case file and background, as well as the slides you observed, what disease does Patient D probably have?

_____

_____

_____

_____

**2.** Why did you make that diagnosis?

_____

_____

_____

_____

_____

**3.** You observed two peripheral blood smears today. Name the pathogens you observed on those slides and the diseases they cause. How are they different microscopically?

_____

_____

_____

_____

_____

# Parasitic Worms

## CASE FILE

On July 19, the director of a children's day camp notified the health department of three campers who had received a diagnosis of a skin condition. The camp included a main building, swimming pools, a volleyball court, a playground with a sandbox, a picnic area, and a beach for boating and swimming. The next day all current campers, their parents, and staff members were notified about the ill children. Parents were asked to look for symptoms of infection on their children, including a snake-shaped (serpiginous) red rash, itching, and pus-containing skin lesions. Twenty-two reports of persons with signs or symptoms of the disease were received. Patients were treated with thiabendazole, mebendazole, albendazole, or ivermectin.

©BSIP SA/Alamy Stock Photo

All of the patients participated in a camp for children ages 2 to 6 years. Although campers from camps were exposed to sand from the beach and the volleyball court, only the younger children were allowed in the playground area, which included a sandbox containing sand that had been placed in the box 2 years previously. Most of the children who became ill did not wear shoes while in the sandbox. Investigators observed cats around the playground and noticed animal feces inside the sandbox.

After analyzing patient symptoms and other information, the health department determined that the sandbox was the source of infection. Two feral cats were removed from the premises by animal control. Health department staff recommended that the sandbox be covered with a tarp when not in use and that the sand be changed regularly. You will be looking at the cause of this infection and others.

## LEARNING OUTCOMES

At the completion of this exercise, students should be able to

- describe the major parasitic flatworms and roundworms, the diseases they cause, and their transmission or life cycles.
- be able to recognize the major parasitic flatworms and roundworms in tissue specimens or other samples.

## Background

Parasitic worms **(helminths)** are common in many areas of the world, infecting millions and in some cases billions of people. The two major categories of parasitic helminths are flatworms and roundworms. **Flatworms** (phylum Platyhelminthes) are among the most primitive of all animals. They have no digestive system at all or an incomplete one with only one opening. Flatworms are hermaphroditic (monecious), lacking separate male and female forms. Some, such as the marine flatworms and *Planaria*, are predators and some (such as tapeworms) are parasites. The **roundworms** (phylum Nematoda) are more complex than flatworms. Unlike the flatworms, they have separate males and females (are dieocious). They have a complete digestive tract with two openings and many other advances in their internal structure. Both roundworms and flatworms use eggs for reproduction. A common strategy in primitive animals that reproduce by means of eggs is to produce huge numbers because so few will survive to become new adults. This means that people infected with parasitic worms will often have the worm producing huge numbers of eggs inside their body. Because so many eggs are produced and released into the feces, examination of **fecal material** can be useful for detection and diagnosis of worm infections.

Flatworms include two major groups of parasites: flukes and tapeworms. **Flukes** typically have complex life cycles with multiple stages and multiple hosts. Typically, juvenile flukes live in invertebrate animals and adults live in vertebrate animals such as fish, sheep, or humans. They have flat, oval bodies with two suckers.

Blood flukes cause the disease schistosomiasis; other disease-causing flukes are the liver fluke and the lung fluke *Paragonimus*. The fluke we will be looking at is the **Chinese liver fluke,** *Clonorchis sinensis* **(figure 30.1),** which has a life cycle that includes humans, snails, and freshwater fish. Humans are usually infected when they eat uncooked or undercooked fish.

**Tapeworms** are intestinal parasites. They are the best known of the flatworms. Their body is made up of two basic parts: the end where they attach to the intestine, known as the **scolex,** and a series of body sections known as **proglottids (figure 30.2).** The scolex is the point of attachment

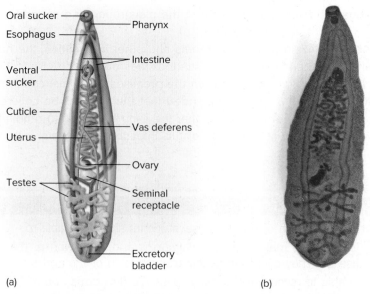

(a)

(b)

**Figure 30.1 The liver fluke *Clonorchis*. (a)** Structures. **(b)** Microscopic appearance. (Magnification: 40×) (b) ©Steven Obenauf

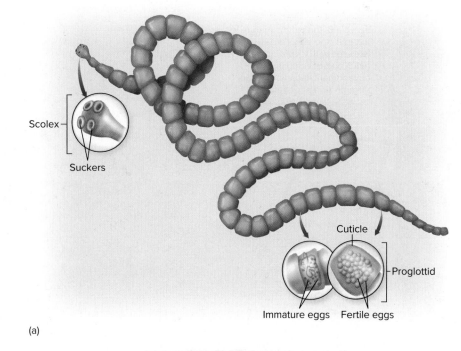

(a)

(b)  (c)  (d)

**Figure 30.2** The tapeworm, *Taenia*. **(a)** Typical tapeworm slide shows the location of sections—observe the two at the arrows. **(c)** *Taenia* scolex (40×). **(d)** *Taenia* proglottids (40×). (b-d) ©Steven Obenauf

to the intestine of the host. It has hooks and suction discs to accomplish that. The structure of the scolex can be used to identify the species of tapeworm a patient is infected with. Proglottids contain the male and female reproductive organs. Tapeworms lack a digestive system and therefore absorb nutrients through their surface. Humans typically get tapeworms by eating undercooked or uncooked beef, pork, or fish containing larvae. Infections with the pork tapeworm cause the most serious symptoms, including siezures. Cases are seen each year in the United States and Latin America.

The **roundworms** (phylum Nematoda) include many very common intestinal parasites, such as *Ascaris*, hookworms, and pinworms. Less commonly, other areas of the body are infected by roundworms such as the filarial worms that cause elephantiasis (in the lymphatics) or canine heartworms.

One example you will observe are the **hookworms.** They suck blood from the intestinal wall. It is estimated that worldwide, one out of every five people have hookworms living inside them. There are two genera of hookworms: *Necator* and *Ancylostoma*. They do not enter the body in food or drink as most intestinal parasites do. Instead, the infective **filariform larvae** of the worms, which are found in soil containing feces, penetrate the skin, usually through the feet (wearing shoes prevents this). They are

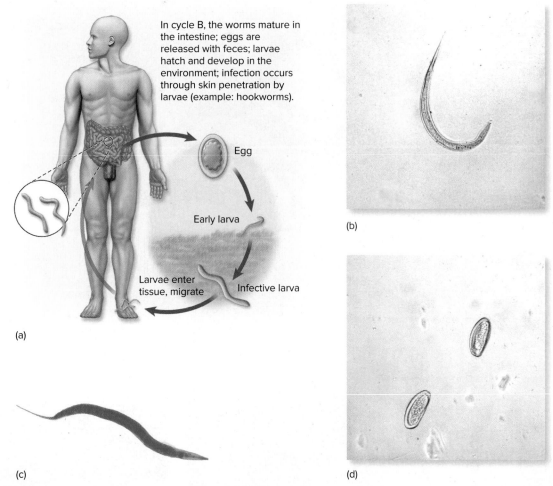

In cycle B, the worms mature in the intestine; eggs are released with feces; larvae hatch and develop in the environment; infection occurs through skin penetration by larvae (example: hookworms).

Egg

Early larva

Infective larva

Larvae enter tissue, migrate

(a)

(b)

(c)

(d)

**Figure 30.3 Roundworms. (a)** Hookworm life cycle; **(b)** hookworm larva (100×); **(c)** pinworm adult (40×); **(d)** pinworm eggs (400×). (b-d) ©Steven Obenauf

carried in the bloodstream to the lungs, where they are coughed up and swallowed, finally reaching the intestine **(figure 30.3a, b).** They then attach to the intestinal wall. The eggs of the worm are shed in feces and then develop into the infective larvae. Humans, dogs, and cats are infected by different hookworms. When dog or cat hookworm larvae penetrate human skin, they can't get past the deeper layers of the skin and so burrow around under the skin. This causes itchy snake-shaped skin lesions where the worm has tunneled beneath the skin. You will be looking at a slide of the infective larvae.

**Pinworms** are another example of intestinal roundworms. They are the most common worm infection in the United States. The adult male and female worms **(figure 30.3c)** live in the large intestine. The female exits the anus at night and lays her eggs **(figure 30.3d)** on the skin in the perianal area. This in turn causes local itching. Humans are infected when they ingest the eggs **(figure 30.4).** Diagnosis often involves microscopic examination of tape placed on the perianal skin.

**Trichinosis** is caused by *Trichinella spiralis*. It is a disease of animals and is transmitted when an infected animal is eaten by another animal (typically a carnivore or a scavenger). The cysts of the worm are found in the skeletal muscle of infected animals. Eating undercooked pork is the most common way humans contract this disease. After consumption, the cysts in the meat release larvae, which leave the intestine and migrate to and encyst in the muscle tissue of the new host. Diagnosis typically involves examination of muscle biopsies **(figure 30.5)** or serological tests.

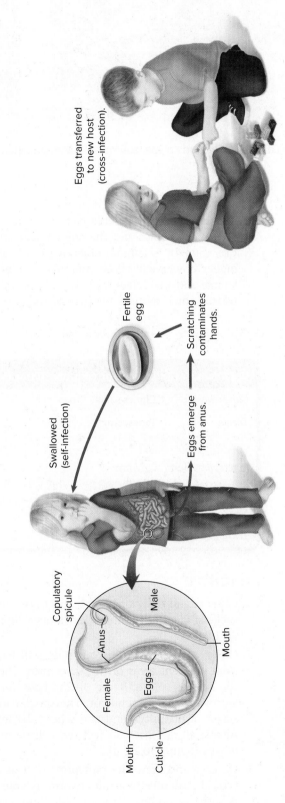

**Figure 30.4** Life cycle of pinworms.

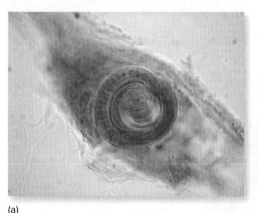
(a)

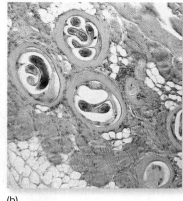
(b)

(c)

**Figure 30.5** Various views of *Trichinella* cysts in sections of skeletal muscle. Magnifications: **(a)** 400×, **(b)** 100×, **(c)** 200×. (a-c) ©Steven Obenauf

### ✅ Tips for Success

- It is important to observe each slide at the correct magnification. *Clonorchis, Taenia,* and the adult *Enterobius* are large specimens: Use your **4×** (scanning) objective (40× total magnification). Using a higher power will allow you to see only a small portion of the animal and may crack the coverslip on your slide as these are usually thick mounts because of the size of the worms.

### Organisms (Prepared Slides)

| Organism | Disease or common names | Sample |
|----------|-------------------------|--------|
| *Clonorchis* | Chinese liver fluke | Adult, whole mount |
| *Taenia* | Tapeworm | Adult, whole mount |
| | | Scolex and proglottids |
| *Enterobius* adult | Pinworm | Adult, whole mount |
| *Enterobius* eggs | Pinworm | Perianal tape lift |
| *Necator* | Hookworm | Infective larvae |
| *Trichinella* | Trichinosis | Cysts in muscle tissue |

### Procedure

1. Observe and draw or photograph *Clonorchis* at 40×. Locate the two suckers. Identify the male and female reproductive parts (figure 30.1*a*).

2. Observe *Taenia* at 40×. Your slide will have several pieces on it from different areas of the tapeworm. The scolex is usually part of the smallest piece (figure 30.2*b*). Look for the structures that the worm uses for attachment. The proglottid pieces will be at different stages of maturity. Find one where the internal structures are visible. Almost all of what you will see will be male and female reproductive organs (figure 30.2*a, d*).

3. Observe and draw the pinworm adult at 40×. It will show the typical structure of a roundworm. Is your animal a male or female? If both are present on your slide, how can you tell them apart?

4. The pinworm eggs will be abundant and widespread on the slide, but you will have to find them in between the other material. Find

them at 40× and observe them at 100× or 400×. Look for the thick outer shell that protects the worm inside. The eggs are clear. You may need to close your condenser (decrease the light) to see the eggs better. Draw what you observe.

5. There are usually only one or two hookworm larvae on the whole slide, so look around with low power (40×) first to find them. Once you have located the larvae, switch up to 100× or 400× to see them better. They may be clear or stained. How do these larvae structurally differ from the tapeworms and flukes? Draw your specimen.

6. Observe and draw *Trichinella* at either 100× or 400×. The specimen you will look at will be muscle tissue. Look for the worms embedded inside it. Depending on how the slice was made, you may need to look around for a while to find a larval worm that is easy to see.

7. Observe and draw any other additional slides your professor may assign.

# NOTES

216

Name _____

Date _____

# Parasitic Worms

## Your Results and Observations

Draw or photograph with your mobile device the slides you observed. Label the parts of the your professor asks you to identify.

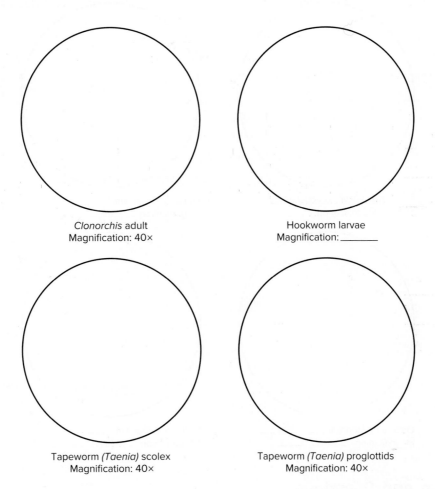

*Clonorchis* adult
Magnification: 40×

Hookworm larvae
Magnification: _____

Tapeworm *(Taenia)* scolex
Magnification: 40×

Tapeworm *(Taenia)* proglottids
Magnification: 40×

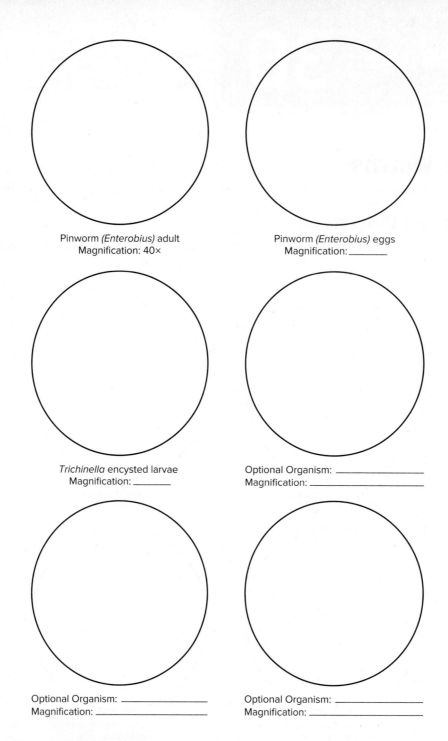

Pinworm *(Enterobius)* adult
Magnification: 40×

Pinworm *(Enterobius)* eggs
Magnification: _____

*Trichinella* encysted larvae
Magnification: _____

Optional Organism: _____
Magnification: _____

Optional Organism: _____
Magnification: _____

Optional Organism: _____
Magnification: _____

## Interpretation and Questions

**1.** With which parasite are the day camp participants infected? What signs or symptoms led you to that diagnosis?

_____

_____

_____

_____

EXERCISE

# 31

# Transmission: Vectors of Disease

## CASE FILE

There are over 40 genera and thousands of species of mosquitoes. Mosquitoes in the genus Aedes can transmit dengue, yellow fever, chikungunya, and Zika viruses. They can breed in and around human homes and bite at any time of the day. Work is under way to develop traps that will selectively attract and capture or kill female Aedes mosquitoes, reducing the use of pesticides. One type of trap, known as an ovitrap, was tested in several communities in Puerto Rico for its possible impact on the numbers of this mosquito. Chikungunya virus had recently appeared in these communities after having never been seen there before. In intervention communities, three traps that killed mosquitoes were put in each home and surveillance traps that only monitored the number of mosquitoes were randomly placed in and around homes. In nonintervention (control) communities, only surveillance traps were used. People living in the houses were questioned about their disease history and samples were drawn for antibody testing by ELISA/EIA to show exposure (or no antibody production, indicating a lack of exposure) to the chikungunya virus. Monitoring showed differences in the average densities of *Aedes* mosquitoes in the communities. Table 31.1 shows the data from those tests.

Source: James Gathany/ Centers for Disease Control and Prevention

## LEARNING OUTCOMES

At the completion of this exercise, students should be able to
- describe the organisms that can act as vectors and the diseases associated with each.
- understand the role that vectors play in microbial disease.

### Table 31.1*

| Community and treatment | Average counts of Aedes per monitor trap | IgG Anti-chikungunya positive residents |
|---|---|---|
| No ovitraps | 11.6 | |
| Community A | | 40.8% |
| Community B | | 55.1% |
| Ovitraps | 1.1 | |
| Community C | | 18.8% |
| Community D | | 28.4% |

*Data and other information from MMWR Vol. 65/No. 18.

**Figure 31.1** *Anopheles* mosquito. ©Steven D. Obenauf

# Background

Infectious diseases can be transmitted in a variety of ways and routes, including direct and indirect contact, respiratory droplets, food and water, and vectors. Vectors are **living organisms that transmit disease** either directly by biting (biological vector) or indirectly (mechanical vector) by carrying disease organisms on or in their body. They belong to the group of animals known as arthropods, which include arachnids (such as spiders, scorpions, ticks, and mites), crustaceans (such as shrimp and crabs), and insects. Many vectors are insects; some are other types of arthropods. The **mosquito** is the best-known insect vector and transmits many important diseases. The *Anopheles* mosquito **(figure 31.1)** is the vector for malaria, the cause of which you observed in exercise 29. The malaria parasite spends part of its life cycle in the digestive tract of the mosquito. In addition to protozoa, mosquitoes can also serve as vectors for parasitic worms such as the filarial worms that cause elephantiasis. Other insect vectors include fleas **(figure 31.2),** deer flies, kissing bugs, and tsetse flies, vectors for *Trypanisoma*, which you also observed in exercise 29. The best-known arachnid vector is the **tick.** Lyme disease is caused by the bacterium *Borrelia* and transmitted by certain species of ticks. The tick transfers the bacteria between different mammal hosts, and *Borrelia* even lives inside of the tick part of the time **(figure 31.3).** For some diseases, where there may not be a vaccine for prevention or drug for treatment, controlling the vector population is the main method of dealing with the disease. This principle was illustrated in the Case File. Vectors can be involved in the process of diagnosing diseases as well.

**Figure 31.2** *Ctenocephalides* flea. ©Steven D. Obenauf

## ✔ Tips for Success

- All of these organisms are large by comparison to microbes. Observe all of these slides with the **lowest power on your microscope** (your 4× objective). Using a higher power will mean you won't be able to see the whole organism and may crack the coverslip on your slide, as these are usually thick mounts because of the size of the arthropods **(figure 31.4).**

## Organisms (Prepared Slides)

| Organism | Diseases it can transmit |
| --- | --- |
| • Mosquito | • Malaria, arboviral encephalitis, yellow fever, dengue fever, elephantiasis, Zika |
| • Tick | • Lyme disease, Rocky Mountain spotted fever, tularemia |
| • Flea | • Plague, typhus |

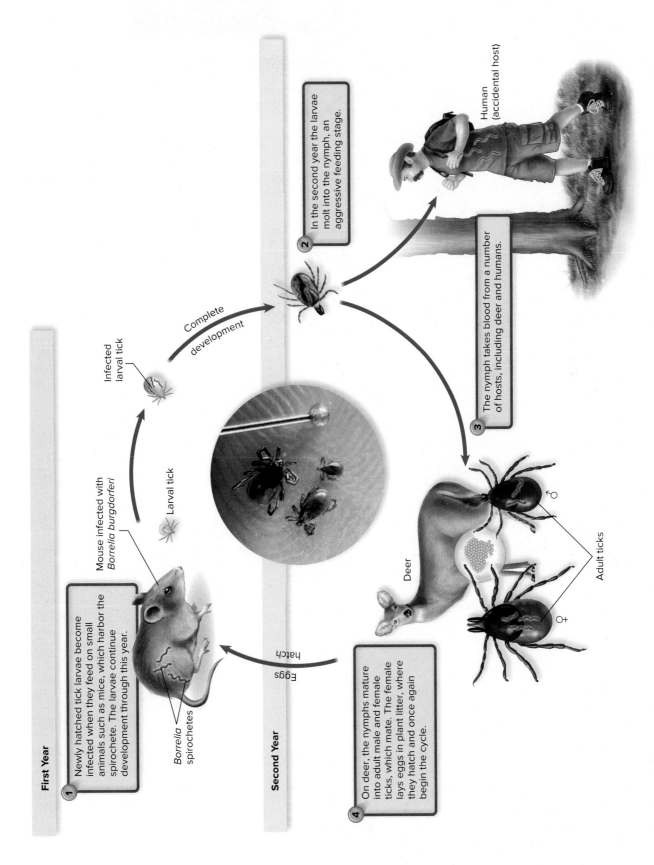

**Figure 31.3** Ticks and the life cycle of Lyme disease. ©Scott Camazine/Science Source

**First Year**

1. Newly hatched tick larvae become infected when they feed on small animals such as mice, which harbor the spirochete. The larvae continue development through this year.

Mouse infected with *Borrelia burgdorferi*

*Borrelia* spirochetes

Larval tick

Infected larval tick

Complete development

**Second Year**

Eggs hatch

2. In the second year the larvae molt into the nymph, an aggressive feeding stage.

3. The nymph takes blood from a number of hosts, including deer and humans.

Human (accidental host)

Deer

Adult ticks

♂

♀

4. On deer, the nymphs mature into adult male and female ticks, which mate. The female lays eggs in plant litter, where they hatch and once again begin the cycle.

**Figure 31.4** Mosquito slide with thick mount and cracked coverslip from using wrong objective lens and not focusing carefully. ©Steven D. Obenauf

## Procedure

**1.** Observe all of these slides with your 4× objective or a dissecting microscope. Draw the organisms you observe.

Name _____

Date _____

# Transmission: Vectors of Disease

## Your Results and Observations

Draw or photograph with your mobile device the organisms you observed.

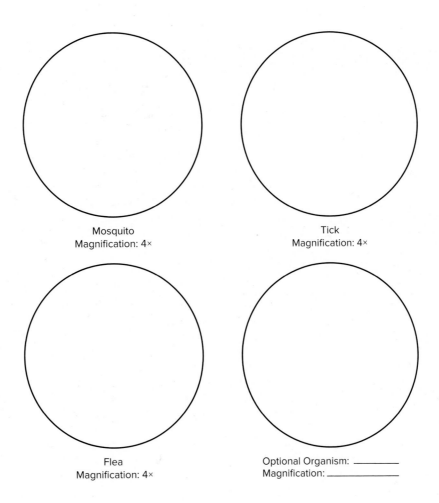

Mosquito
Magnification: 4×

Tick
Magnification: 4×

Flea
Magnification: 4×

Optional Organism: _____
Magnification: _____

## Interpretation and Questions

**1.** Based on the data in table 31.1, did the ovitraps prevent the transmission of chikungunya? If yes, how?

_____

_____

_____

_____

**2.** Based on the life cycle illustrated in figure 31.3, if there are mice but no deer in an area, would humans be at risk for Lyme disease? Why?

_____

_____

_____

_____

# Gel Immunoprecipitation (Immunodiffusion)

## CASE FILE

One of the potential complications of rubella in newborn infants is congenital rubella syndrome, which can cause heart defects, deafness, and many other problems. The virus passes through the placenta from mother to child if the mother is infected, so it is important that women of childbearing age be protected against this disease. In some women, the rubella vaccine they received as a child loses its effectiveness. A rubella titer test can be performed as part of prenatal care of pregnant women. It is sometimes used to screen women of childbearing age before the first pregnancy. If the patient has little or no immunity, the titer will be 1:8 or less (between 1:1 and 1:8). Women without immunity should be revaccinated. Your job is to do a titer test to find out if your patient (Patient T) is protected against rubella.

Source: Centers for Disease Control

## LEARNING OUTCOMES

At the completion of this exercise, students should be able to

- understand precipitation and how it is used diagnostically.
- perform and interpret a gel immunoprecipitation test.

## Background

Antibodies are proteins produced by the immune system that specifically bind to materials the immune system recognizes (known as antigens). The amount of antibody found in the blood to a specific antigen increases significantly after the immune system is exposed to the antigen either naturally or by vaccination. This is part of what is known as immunological memory. As a result of memory that vaccination creates, diseases that once caused thousands or tens of thousands of deaths each year are now routinely prevented. An unfortunate effect of the current attacks on vaccines by some individuals is a recent spike in dangerous diseases, such as whooping cough and measles, that had barely been seen in decades. **Precipitation reactions** happen when antibodies and soluble antigens bind to each other and form insoluble complexes, which precipitate from solution **(figure 32.1).** In the body, they are one of the many mechanisms the immune system uses to eliminate foreign cells or materials. They can also be carried out in the laboratory to visualize the binding of antibody and antigen as a diagnostic

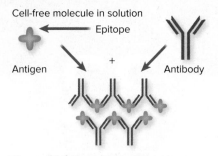

**Figure 32.1** Precipitation reactions.

test. In **immunoprecipitation testing,** precipitation reactions are done in agarose (a purified form of agar). The antigen and antibodies each diffuse out of the wells they are placed in (which is why the reaction is also called immunodiffusion). If the antibody can recognize and bind to the antigen, and they are in the correct amounts, a visible precipitate forms in the agarose between the antigen and antibody wells.

**Titer** is a measure of the amount of antibody in a person's serum. The higher the titer is, the more protection a person has against a particular disease. Titer can be measured by diluting the person's serum (1:2, 1:4, 1:8, and so on) and seeing which dilutions contain enough antibody to react. In this testing setup, *titer* is defined as the last or highest dilution that still gives a positive result (in this exercise, produces a precipitate).

## ✅ Tips for Success

- When you are loading the antigen and antibodies into the agarose plate, don't push the tip of your micropipette into the wells, or they can overflow. Hold it just above the opening. This will take a steady hand.
- When you read your results during the second period, holding the plate above you with lighting from the side against a dark background works best.

## Materials (per Team of Two or Four)

- 1 agarose plate **(figure 32.2a)** (plates are 0.8% agarose in buffer prepoured and punched)
- 10 μL micropipettor and tips
- Rubella antigen (on ice or in cold block)
- Patient T serum dilutions (undiluted, 1:2, 1:4, 1:8, 1:16, 1:32) (on ice or in cold block)
- Ziploc bag
- Paper towel
- Sharps container for disposing of micropipette tips
- Parafilm (optional)

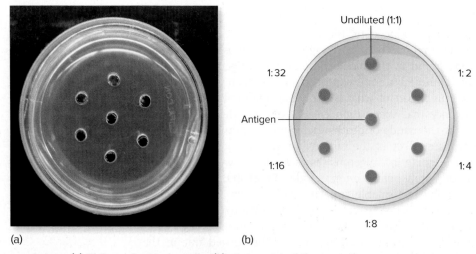

(a)  (b)

**Figure 32.2 Agarose plate. (a)** Plate and sample wells; **(b)** placement of the samples. (a) ©Steven D. Obenauf

## Procedure

### Period 1

1. Label the agarose plate (on the **outer edge** or outside of the wells on the bottom) as shown in the diagram in **figure 32.2b.**

2. Carefully use the micropipettor **(figure 32.3)** to put 10 μL of antigen into the center well and 10 μL of each antibody dilution into the outer wells.

3. Change tips between each sample. Note the antibody dilutions to make sure you put the right ones in the right wells.

4. Discard used pipette tips in the sharps container.

5. Place the immunodiffusion plate on a moist paper towel (to maintain a moist environment and enhance diffusion) inside a plastic Ziploc® bag in the student refrigerator on your class's shelf. Optional: Wrap the edges of the plate with Parafilm.

### Period 2

1. If your plate is cold, let it warm up to room temperature before you look at it.

2. Hold the plate up and observe the area of precipitation caused by the antigen/antibody response **(figure 32.4).** Hold the plate so that the light comes from the side to make the precipitation more visible. How many of the different antibody dilutions show precipitation? Look carefully–some may be faint.

3. Draw or photograph and record your results.

## Results and Interpretation

Look at each dilution, starting with undiluted serum (1:1) and going around to the highest dilution, 1:32. Precipitation will look whitish and cloudy and will usually curve around the outer well **(figure 32.5).** Look carefully– they may be faint and hard to see, especially at higher dilutions. The plate in figure 32.5 shows positive reactions with 1:1, 1:2, and 1:4, and negative reactions with 1:8, 1:16, and 1:32. Since 1:4 is the highest dilution with a positive reaction, this plate shows a titer of 1:4 (this can also be expressed as 4 or 1/4). If there had been positive reactions at 1:8 and 1:16 dilutions, the titer would have been 1:16, which is a higher titer than 1:4.

**Figure 32.3** A 10 μL micropipettor.
©Steven D. Obenauf

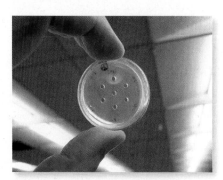

**Figure 32.4** Holding up the plate to observe precipitation.
©Steven D. Obenauf

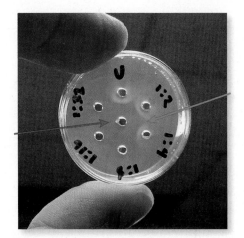

**Figure 32.5 Precipitation reactions in an agarose plate.** In this plate, reactions can be seen in the undiluted, 1:2, and 1:4 wells. ©Steven D. Obenauf

# NOTES

# Gel Immunoprecipitation (Immunodiffusion)

## Your Results and Observations

Draw or photograph your results.

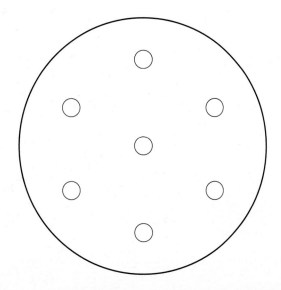

| Dilution | Positive or negative |
|----------|---------------------|
| Undiluted | |
| 1:2 | |
| 1:4 | |
| 1:8 | |
| 1:16 | |
| 1:32 | |

## Interpretation and Questions

1. Based on your results, what is your patient's titer?

_____

_____

_____

_____

**2.** Does Patient T need another rubella vaccination? What is the basis for your recommendation?

_____

_____

_____

_____

_____

_____

_____

# Passive (Indirect) Agglutination

## CASE FILE

Patient C, a 68-year-old woman, was diagnosed with rheumatoid arthritis (RA) 11 years ago. One year after diagnosis, she began treatment with a nonsteroidal anti-inflammatory drug (NSAID). Three years ago, she retired and started backyard gardening in her Idaho community. She discontinued her treatment at that time. One year ago, her arthritis symptoms worsened. Resumption of NSAID therapy had not helped, but the addition of Humira®, a biological response modifier and disease-modifying anti-rheumatic drug, seemed to reduce her symptoms. The patient was evaluated to determine if more aggressive drug therapy was needed.

©Science Photo Library/ Alamy Stock Photo RF

On physical examination, her elbows, wrists, and metacarpophalangeal joints showed mild swelling and tenderness. Both knees had effusions (extra fluid) and were painful on flexion. She had no subcutaneous nodules (a common symptom of RA) and no other signs of systemic illness. X rays of the affected joints were ordered. In order to find out if her current treatment was sufficient, a test to detect elevated C-reactive protein (CRP) levels was also ordered to see if significant levels of inflammation were still present in Patient C.

### LEARNING OUTCOMES

At the completion of this exercise, students should be able to
- understand agglutination and how it is used diagnostically.
- perform and interpret a passive agglutination test.
- describe the clinical importance of C-reactive protein levels.

## Background

**Agglutination** reactions happen when antibodies bind to particulate antigens or cells such as bacteria. These interactions are visible as an agglutinate or clumping of cells **(figure 33.1)**. In the body, they are one of the many mechanisms the immune system uses to eliminate foreign cells or materials. Immune reactions are also frequently used as part of a variety of **diagnostic testing methods.** All immunologically based diagnostic testing methods use something to make antigen-antibody reactions detectable. Examples include precipitation, as we saw in exercise 32, addition of fluorescent dyes, color-generating reactions (exercise 34), and clumping of cells or beads to make the antigen-antibody reactions easy to detect or quantify, as you will

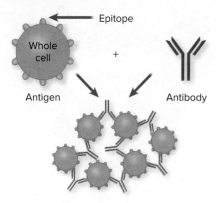

Microscopic appearance of clumps

**Figure 33.1** Antibodies, cells, and agglutination reactions.

see in this exercise. **Passive agglutination** (also known as indirect agglutination) assays utilize beads to make the reaction visible. They are very widely used due to their flexibility, speed, and ease of use. Depending on what the test is designed to look for, either antigens or antibodies can be attached to the beads. When antigen has been attached to the beads, they will agglutinate if the appropriate antibody is in the patient's serum. The presence of antibodies to a pathogen is often used to diagnose the disease associated with it. When antibody has been attached to the beads, they will agglutinate if the target antigen is present in the patient's sample.

**C-reactive protein (CRP)** is a serum protein (of a class known as acute-phase proteins) that is elevated in inflammatory conditions such as bacterial infections and **rheumatoid arthritis.** Rheumatoid arthritis is an inflammatory autoimmune disease whose effects are often observed in the hands. Osteoarthritis is a degenerative, or "wear-and-tear," condition frequently impacting knees and hips. CRP has been shown to be an early indicator of inflammation, as CRP levels tend to increase before rises in antibody titer and erythrocyte sedimentation rate (ESR). CRP levels also drop faster than ESR when inflammation subsides, so they can be used to determine how well treatment is working. Elevated CRP levels may also indicate risk of cardiovascular disease just as much as cholesterol levels..

The latex agglutination slide test reagent you will use consists of **latex beads** coated with antibodies against human CRP. When a serum specimen containing CRP is mixed with the latex reagent, binding of CRP to the beads will occur, resulting in visible agglutination of the latex beads. If the specimen is negative for CRP, no agglutination (a uniformly turbid suspension) will result.

## ✓ Tips for Success

- Make sure all of the members of your team are present when you do this test. The reaction happens quickly, so they will need to be there to see it.
- Hold the dropper bottles vertically and close to the test slide when you put on the samples. The drops should be fairly small. Large drops can sometimes interfere with the reaction.
- Put the drops for the controls and patient sample off-center, so that you can put the latex reagent drop next to them rather than on top of them.

## Materials (per Team of Two or Four)

- IMMUNEX CRP latex reagent (on ice or in cold block)
- Positive control, negative control (on ice or in cold block)
- Patient C sample (on ice or in cold block)
- Test slide
- Toothpicks
- Beaker of disinfectant
- Sharps container

## Procedure

1. Obtain a test slide (it is black with white rings). You will be using three of the six wells on the slide. Obtain three toothpicks and bottles of positive control, negative control, patient's sample, and latex reagent **(figure 33.2).**

2. Place one **small** drop each of the positive control, negative control, and patient's sample in the appropriate circle on the slide (follow the diagram in **figure 33.3).** WEAR GLOVES WHEN HANDLING SERUM SAMPLES.

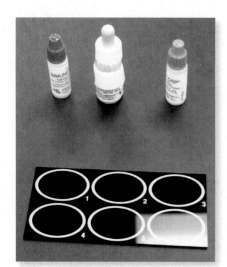

**Figure 33.2** Positive and negative controls, latex reagent, and test slide.
©Steven D. Obenauf

3. Shake the latex reagent prior to use. Add one drop of latex reagent to each circle next to the drop that is already there. To avoid contaminating the reagent bottle, be careful not to touch the drops already on the slide.

4. Using a new stirrer (toothpick) for each ring, mix thoroughly and spread over the entire circle.

5. Rotate the slide **slowly and gently** in a circular, horizontal motion while the reagents react **(figure 33.4).** Stop when the positive control has fully agglutinated. (You will see white specks starting to form around the outer edge.)

6. Observe each circle at around 2 minutes for agglutination—make sure that all of the members of your group are present. Positive results show visible agglutination. Negative results will appear uniformly turbid.

7. Record your results and dispose of used stirrers in the sharps container and used test slides in the disinfectant beaker.

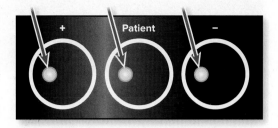

**Figure 33.3** Locations for adding the controls and patient's sample to the test slide (left) and adding the latex reagent (right).

## Results and Interpretation

Negative reactions will be uniformly cloudy, as seen in the negative controls (well 6 of **figure 33.5a** and well 3 of **figure 33.5b**) and in the patient's sample (well 2) of slide (b). Positive reactions are usually strongest around the edges of the circle. Clumping of the latex beads will show as white specks against the black slide, as seen in the positive controls (well 4 of figure 33.5a and well 1 of figure 33.5b) and the patient's sample (well 5) of slide (a).

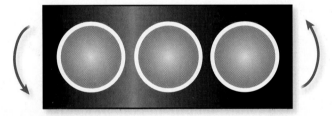

**Figure 33.4** Rotating the slide.

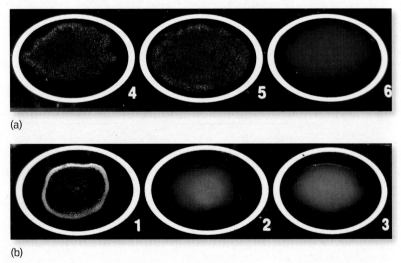

(a)

(b)

**Figure 33.5** Passive agglutination slides with **(a)** positive and **(b)** negative reactions in the patient's (center) sample well. (both) ©Steven D. Obenauf

# NOTES

# Passive (Indirect) Agglutination

## Your Results and Observations

Photograph your results or draw them on the following diagram.

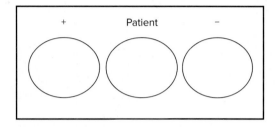

+     –

Record the reaction in Patient C's sample.  [ ]    [ ]

## Interpretation and Questions

**1.** Do you think Patient C's medications need to be changed? Why or why not?

_____

_____

_____

_____

**2.** Why is passive agglutination a useful diagnostic tool?

_____

_____

_____

_____

**3.** What is CRP and why do we test for it?

_____

_____

_____

_____

# Rapid Immunoassays

## CASE FILE

Your patient, a 4-year-old boy (Patient A), was complaining that his throat hurt. When speaking to him, his voice seemed hoarse. He has a history of allergic rhinitis (allergies) triggered by pollen and was treated as needed with antihistamines or occasional intranasal corticosteroids. He vomited after coming home and was brought to the pediatrician for an examination. He was found to have a temperature of 39°C (102.2°F), a red throat, and some swelling of the lymph nodes in his neck. A whitish exudate was seen on his tonsils. On auscultation and palpation, his lungs were found to be clear. Diagnostic tests were ordered. A throat swab was taken for culture on a blood agar plate (exercise 14) and a rapid immunoassay for Group A *Streptococcus*. Blood was drawn for a Monospot™ test for infectious mononucleosis and a differential white blood cell (WBC) count. An antibiotic sensitivity test (exercise 25) was also ordered.

Source: Melissa Brower/ Centers for Disease Control and Prevention (CDC)

## LEARNING OUTCOMES

At the completion of this exercise, students should be able to

- describe the principle involved in how immunoassays work.
- read and interpret the results from a lateral flow device test.

## Background

**Immunoassays** are quick and accurate tests that can be used in the laboratory, in a health care setting, or even at home to detect specific molecules such as a microbial protein, hormone, or drug. Immunoassays rely on the ability of an antibody to bind specifically to a target molecule. The two assays we will describe here, Enzyme immunoassays **(EIAs)** and lateral flow devices, have differences but also some things in common. They both utilize a patient sample, an antibody against what is being tested for, and something that creates color. A third immunoassay, the radioimmunoassay, uses radioactivity to detect the reaction. EIAs are usually conducted in a flat plate with multiple small wells. There are many different types of EIAs, but they all involve an antibody molecule with an enzyme attached. The most widely used enzymes are alkaline phosphatase and horseradish peroxidase. The addition of a substrate for the enzyme allows a determination if a reaction has taken place. The substrate for the enzyme

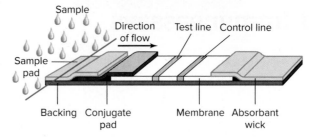

Figure 34.1 Typical lateral flow device. ©Steven D. Obenauf

is typically a colorless molecule, known as a chromogen, that is converted into a colored end product. EIAs are read by looking for color. EIAs can give a "yes-or-no" answer or be quantitative. EIAs are extremely sensitive, allowing antigen to be quantified in the nanogram per milliliter range. One of the most commonly discussed uses for EIA testing is checking blood for antibodies to HIV.

With lateral flow devices **(LFDs)**, a solution containing the sample is applied to the end of a test strip of some sort and flows by capillary action along the strip **(figure 34.1)**. The test strip is essentially a piece of paper, with the antibody and red or blue colored beads attached in specific locations (lines) that will generate colored bands if a reaction takes place. The fluid in the sample actually hydrates the reagents, which are present in a dried state in the strip. Antibody coated with labels such as gold is dried on the strip, where it interacts in the conjugate area with the added sample as it flows laterally along the membrane. The antibody will bind to (capture) the antigen (if it is present) as it flows by (tests can be set up to detect antibodies as well). The antibody-antigen complexes then flow over a second line; this line will retain the beads that have bound antigen. A third control line binds any beads. If the color only develops at the control line, the test is negative. These LFD tests are quick and easy to perform, making them popular for in-office or in-home use. In-home pregnancy tests are the best-known example of a lateral flow test. However, LFDs are not as sensitive nor can they be used for quantitation.

©Steven D. Obenauf

©Steven D. Obenauf

©Steven D. Obenauf

## ✔ Tips for Success

- Do not remove testing dipsticks from the foil pouch until ready to perform the assay.
- To avoid cross-contamination, do not allow the tip of the reagent A or B bottles to come in contact with the sample swab.

## Materials (per Team of Two or Four)

- Patient A sample (on plate)
- 1 Quidel™ QuickVue™ strep A test dipstick
- Extraction reagent A (4 M sodium nitrite)
- Extraction reagent B (0.2 M acetic acid)
- Sterile swab
- Test tube

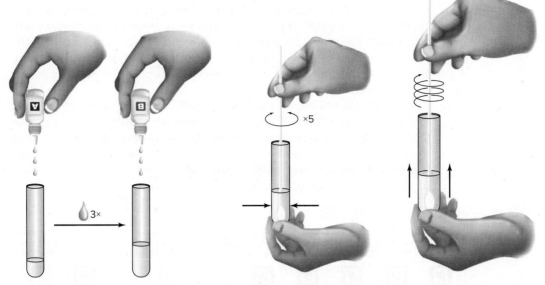

**Figure 34.2** **(a)** Adding reagent A and B; **(b)** squeeze and rotate; **(c)** squeeze and remove.
©Steven D. Obenauf

## Procedure

1. Wear **gloves** while performing this procedure. Add 3 drops of reagent A to the tube **(figure 34.2a).** When adding drops, hold the reagent bottle **vertically,** so that a complete drop forms.

2. Add 3 drops of reagent B to the tube. The liquid in the tube should turn green after adding the reagent B (see the bottom of the tube in figure 34.3).

3. Lightly touch a colony on the Patient A plate with the sterile swab.

4. Immediately add the Patient A swab sample to the tube. Squeeze, so that the swab head is compressed against the side of the tube to extract the bacteria into the reagents. Rotate the swab a minimum of five times **(figure 34.2b).**

5. Keep the swab in the tube for 1 minute.

6. Express/squeeze out all the liquid from the swab against the inside of the tube. Squeeze the swab firmly as it is removed from the tube **(figure 34.2c).**

7. Discard the swab in a biohazard container.

8. Remove the test dipstick from the foil pouch. Place the dipstick in the tube with the arrows on the dipstick pointing down **(figure 34.3).**

9. Do not handle or move the dipstick until the test is complete and ready for reading.

10. Read and record the result at 5 minutes. (Some positive results may appear sooner.)

11. When all members of the team have seen and recorded the results, discard the dipstick in a biohazard container.

## Results and interpretation

Most cases of pharyngitis are caused by viruses, including the one that causes infectious mononucleosis. These cannot be treated with antibiotics. Group A *Streptococcus* (*Streptococcus pyogenes*) is the cause of pharyngitis in

**Figure 34.3** Dipstick in tube prior to reading results. ©Steven D. Obenauf

roughly 25% of cases in children and 10% of cases in adults. Other bacteria can also cause pharyngitis, including *Haemophilus, Corynebacterium, Neisseria, Staphylococcus, Mycoplasma,* and other species of *Streptococcus.* Potential serious complications of *S. pyogenes* include rheumatic fever and abscesses. The dipsticks in the test you have performed are coated with rabbit polyclonal anti–Group A *Streptococcus.* This will bind to *Streptococcus pyogenes* in the sample and lead to a positive reaction.

Any pink-to-red line in the "Test Region" area along with any shade of a blue line in the "Control Region" is a **positive result** for the detection of Group A *Streptococcus antigen* **(figure 34.4a).** A blue line in the "Control Region" area and no pink line in the "Test Region" area is a **negative result** **(figure 34.4b).** The test result is **invalid** if a blue line in the "Control Region" is not visible at 5 minutes **(figure 34.4c).**

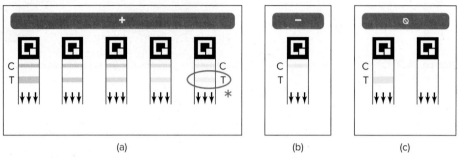

(a)           (b)           (c)

**Figure 34.4** **(a)** Positive result; **(b)** negative result; **(c)** invalid result. Even if you see a very faint pink test line (and a blue control line), this is a positive result.
Credit: Quidel Corporation, San Diego, California

# Rapid Immunoassays

## Results and Observations

Record and draw (or photograph with your mobile device) your results:

|  | Line present (yes/no) |
|---|---|
| Control Region |  |
| Test Region |  |

## Interpretation and Questions

**1.** Do the results indicate that Patient A's symptoms are caused by Group A *Streptococcus*?

_____

_____

_____

_____

**2.** Why was a Monospot™ test ordered?

_____

_____

_____

_____

**3.** In a clinical setting, what type of patient sample would be used to perform this test?

_____

_____

_____

_____

# NOTES

# Hemagglutinin Inhibition Assay

## CASE FILE

A 57-year-old man visited his primary care physician in late August. He reported that he had a fever; was coughing, sneezing, and experiencing muscle aches; and generally did not feel well. The patient, a local farmer, claimed to have felt fine the morning before but by the afternoon was quite ill. He had gone to bed early and had awoken with a fever, chills, and headache and was unable to open his right eye (crusted shut). The patient's blood pressure and urinalysis were normal. His lungs were clear. The patient's temperature was elevated 39° C (102.2° F). The symptoms and signs, including unilateral conjunctivitis, are typical of viral respiratory infections, such as influenza.

Source: https://www.cdc. gov/flu/professionals/ laboratory/antigenic.htm

The physician, suspecting the patient had the flu, ordered the office laboratory technician to perform a rapid influenza diagnostic test (RIDT). RIDTs are immunoassays used to detect specific strains of influenza A and B viral nucleoprotein antigens. The test can be done quickly and is particularly useful in clinical and institutional settings. The test was performed and the result was negative. False negatives can occur. Additionally, RIDTs vary in their sensitivity to viral strains. Even though the RIDT was negative, the physician was confident the patient had the flu. Because the patient had preexisting conditions (diabetes and had contracted pneumonia twice the previous winter season) and the symptoms had appeared less than 48 hours earlier, the doctor prescribed an antiviral drug (oseltamivir).

The doctor thought that with medication, rest, and time his patient would recover; however, she was concerned for three other reasons. First, this patient contracted flu well before the typical time of onset of seasonal flu for this area; initial cases of flu heralding the onset of flu season usually appear in early October. Because the patient did not report any travel out of the immediate area in the past month, it was unlikely he contracted the flu elsewhere and brought the virus back home and no one else in the home reported being ill. Second, pig and chicken farming are common occupations in this region. These animals are reservoirs for influenza virus from which novel new viruses occasionally jump to human subjects. Third, this is a rural farming community with many lakes, which serve as both sources of irrigation and habitats for waterfowl. It is located along one of the major migratory pathways for geese. Since waterfowl, such as geese, are the primary reservoirs for influenza A viruses, the possibility exists that a novel influenza A virus was transmitted from waterfowl to the patient.

Nasopharyngeal swabs and a blood sample were taken from the patient and sent to the Flu Surveillance Laboratory at the Centers for Disease Control and Prevention (CDC). The Flu Surveillance Laboratory monitors and assesses more than 2,000 influenza viruses a year from around the world as it looks for changes in antigenic properties in circulating viruses. By noting and tracking changes in global viruses, the CDC can better predict which viruses will be the greatest threats for the next flu season.

---

### LEARNING OUTCOMES

At the completion of this exercise, students should be able to
- explain how hemagglutinin inhibition assays are performed.
- describe the basis for classification of influenza A viruses.
- explain titer and hemagglutination.

---

## Background

### Influenza

The flu, or influenza, is often discounted as a minor medical ailment, when, in fact, influenza kills thousands of people in the United States alone every year. The CDC estimates that during the 2015-2016 flu season there were 25 million influenza illnesses, 310,000 influenza-related hospitalizations, and approximately 12,000 related deaths. Worldwide, the World Health Organization estimates, there are up to 5 million severe cases of influenza cases that result in anywhere from a quarter of million to half a million deaths each year. So the flu is certainly not a minor medical ailment. The flu disproportionally impacts the elderly and children under 5. The majority of hospitalizations and deaths are among adults 65 years old and older. Vaccination is effective, but less than 40% of the population gets vaccinated each year.

Influenza viruses belong to the family Orthomyxoviridae. The family is composed of negative-sense single-stranded RNA viruses, including four species of influenza: influenza A, influenza B, influenza C, and influenza D. The host range for influenza A includes humans, pigs, birds, horses, and bats. Geese, ducks, and other wild aquatic birds are considered the natural hosts for influenza A viruses. Influenza A is the most virulent form of influenza for humans. Influenza viruses are named by their serotype or subtype. Influenza A virus serotypes are named by the type of hemagglutinin (H or HA) and neuraminidase (N or NA) found in their viral envelop. Hemagglutinin is a viral envelop protein that binds the viral envelop protein to molecules on the surface of red blood cells, causing agglutination. There are 18 known HA serotypes. Neuraminidase is a viral enzyme that aids the virus in penetration of cells. There are 11 known NA serotypes. When a flu virus is described as H1N1, we know that this is an influenza A virus and its serotype is hemagglutinin 1, neuraminidase 1. Humans are the primary hosts for influenza B. There are fewer serotypes of influenza B, because there is less genetic drift in this virus. Influenza B causes a milder respiratory disease and is most commonly observed in children. Influenza C has a host range of pigs, dogs, and humans. Although influenza C has been associated with localized epidemics, it is extremely rare in humans. Influenza D is restricted to pigs and cattle.

Medical practitioners and researchers, when discussing possible pandemic virus outbreaks, often point to influenza A as an example. All flu pandemics (spread of disease over several countries or continents) that we know of, including the Spanish flu of 1918, the Asian flu of 1957, the 1968 H3N2 flu, and the 2009 H1N1 flu, have been forms of influenza A. The Spanish flu alone was thought to have killed over 1 million people worldwide. Scientists attribute the virulence of this serotype to a novel combination of antigenic properties. They believe that humans had not been exposed to this combination of antigenic determinants before or had limited exposure to this virus previously and therefore had no immunity. The lack of immunity led to high mortality. Mutations in HA and NA genes of influenza A lead to changes in the virus in a process known as **antigenic drift.** Antigenic drift is thought to result from the continuous, ongoing accumulation of point mutations within the genome. These changes can create new versions of HA and NA to which antibodies generated against older viral strains will no longer be effective. Influenza A also undergoes **antigenic shift.** Antigenic shift is a more dramatic episodic occurrence and can occur in several ways, including when the virus jumps between species with or without genetic change (e.g., an avian virus infects a human) or when more than one virus infects a cell and exchanges genetic information. Regardless of the exact mechanism, these two processes, antigenic drift and antigenic shift, make influenza a moving target for the immune system. The constant mutation of the virus and the reality of a globally mobile society suggest that another pandemic could be more than science fiction.

## Hemagglutinin Inhibition Assay (HI)

Hemagglutinin inhibition assays can be used to identify the specific hemagglutinins present in a virus in a sample, which in turn can be used in determining the viral cocktail to be used in creating the next season's flu vaccine. Titer, the concentration of virus in a sample, can also be determined with HI.

For the purposes of this laboratory, HI will be addressed at a very fundamental level. HI is done in a microtiter plate.

Blood cells are large cells. They do not stay suspended in solution, but settle out of solution in 15 to 30 minutes unless the solution is constantly stirred. If blood cells alone were added to the microtiter plate, the cells would settle to the bottom and form a red spot, a pellet, in the bottom of the plate well (row A). Influenza viruses have hemagglutinins, which bind to markers, sialic acid receptors, on the surfaces of cells. If influenza viruses are present in a solution of red blood cells, the viral hemagglutinins and red blood cells interact. This produces a large network of interconnected cells, a lattice or matrix of cells linked by viruses. The process of viruses binding to red blood cells and producing this lattice is called hemagglutination. This large network remains suspended in solution. The solution appears homogenous with evenly dispersed color (row B). HI is performed by combining known viral antigen antibodies with patient serum and then with red blood cells. For example, the patient from the case file is suspected of having influenza A of unknown serotype. The patient's sample containing virus is dispensed into several reaction tubes and incubated with individual antibodies to specific hemagglutinins (H1, H2, H5, H7). After incubating with the antibody, the antibody-virus solution is mixed with red blood cells. If the antibodies bind to the virus blocking the hemagglutinin, then hemagglutination is prevented (row C) and the red blood cells in the mixture naturally settle to the bottom

| Components | Interaction | Microtiter Results |
|---|---|---|
| **A** RBCs | | No reaction |
| **B** Virus RBCs | | Hemagglutination |
| **C** Virus Antibody + RBCs | | Hemagglutination inhibition |

Source: https://www.cdc.gov/flu/professionals/laboratory/antigenic.htm

of the well. For example, suppose the patient has influenza A serotype H5N3. The patient's serum containing virus is added to the reaction tube along with anti-H5 antibody. The antibody binds to the hemagglutinin on the virus, which blocks its ability to bind to red blood cells. When red blood cells are added to the reaction tube, the hemagglutinin is no longer exposed and cannot bind to the sialic acid receptor. Hemagglutination cannot occur and the red blood cells will settle to the bottom of the tube. What would happen if the virus were actually influenza A serotype H1N1 and anti-H5 antibody were added to the sample?

This test is more commonly used to determine the antigenic similarity between viruses by examining the differences in antibody recognition of antigens and the resulting titer. Two or more viruses in circulation are compared to antibodies to prevalent forms of influenza (seasonal flu). The differences in titer responses between the viruses can result from antigenic drift or antigenic shift and provide early evidence of diminishing vaccine efficacy. In the example below, Circulating Virus 1 responds similarly to the previous season's vaccine and so a high level of cross reactivity could be expected. However, Circulating Virus 2 is significantly different in its antigenic properties from the previous season's vaccine virus. Individuals exposed to Circulating Virus 2 would be afforded less protection from the current deployed vaccine.

| | 1:10 | 1:20 | 1:40 | 1:80 | 1:160 | 1:320 | 1:640 | 1:1280 | 1:2560 | 1:5120 | 1:10240 | 1:20480 |
|---|---|---|---|---|---|---|---|---|---|---|---|---|
| Previous season's vaccine virus  A | | | | | | | | | | | | |
| Circulating virus 1 ("like" virus)  B | | | | | | | | | | | | |
| Circulating virus 2 (low reactor)  C | | | | | | | | | | | | |

Source: https://www.cdc.gov/flu/professionals/laboratory/antigenic.htm

## ✅ Tips for Success

- This is a simplified procedure designed to provide you with exposure to this important technique.
- Pipette your volumes carefully.
- Change pipette tips between solutions.
- Mix solutions well, but treat the red blood cell preparation gently. Ruptured red blood cells will produce false results.

## Materials

- V-bottom microtiter plate (12 column)
- 0-100 μL micropipette
- Micropipette tips
- Prepared red blood cells
- Anti-H1 antibodies (diluted)*
- Anti-H5 antibodies (diluted)*
- Patient sample*
- Phosphate buffered saline (PBS)
- *Pre-diluted patient samples or antibody may be provided.

## Procedure

1. Obtain a microtiter plate from the supply table and label the plate as shown.
2. Add 25 μL of PBS to wells 1:20 through 1:20480 in rows A, B, and C.
3. Add 50 μL of the patient sample to well 1:10, rows A, B, and C. Discard the pipette tip.

246

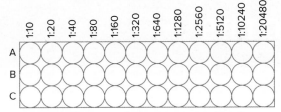

Source: https://www.cdc.gov/flu/professionals/laboratory/
antigenic.htm

## Serial Dilution

**4.** Place a clean tip on the micropipette. Set the pipette to withdraw 25 µL. Mix the contents in well 1:10 in row A by inserting the pipette tip into the well and depressing and releasing the plunger two or three times. Do this gently. After the contents are mixed, remove 25 µL from the 1:10 well and dispense it into well 1:20 (row A). Mix the contents of row A, well 1:20, by pipetting 25 µL of solution back and forth. Remove 25 µL and transfer the volume to the 1:40 well. Mix the solution in this well. Remove 25 µL and transfer the volume to the 1:80 well. Repeat this procedure until you have reached the end of the row. After mixing the 1:20480 well, remove and discard 25 µL of solution. Discard the pipette tip.

**5.** Repeat step 4 for well in row B.

**6.** Place a clean tip on the micropipette. Mix the contents of row C, 1:10 well, by pipetting 25 µL of solution back and forth. Remove 25 µL and transfer the volume to the 1:20 well. Mix the solution in this well. Remove 25 µL and transfer the volume to the 1:40 well. Repeat this procedure until you have reached **1:1280.** After mixing the 1:1280 well, remove and discard 25 µL of solution. Discard the pipette tip.

**7.** Place a clean tip on the micropipette. Add 50 µL of PBS to row C, wells 1:10–1:1280. Add 75 µL of PBS to row C, wells 1:2560–1:2040.

**8.** Add 50 µL of anti-H1 antibody to wells 1:10 through 1:20480 in row A.

**9.** Add 50 µL of anti-H5 antibody to wells 1:10 through 1:20480 in row B.

**10.** Cover the plates and incubate at room temperature for 30 minutes on a shaker. If a shaker is not available, then rotate or agitate the plate to thoroughly mix the contents.

**11.** After the incubation, uncover the plate and **gently** add 50 µL of prepared blood cells to each well in rows A, B, and C.

**12.** Cover the plate and incubate again for 30 minutes.

**13** Observe the plate. Record your results.

# NOTES

Name _____

Date _____

# Hemagglutinin Inhibition Assay

## Your Results and Observations

Record your results. Color the microtiter plate below to reflect the results you obtained.

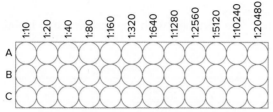

Source: https://www.cdc.gov/flu/professionals/laboratory/antigenic.htm

## Interpretation and Questions

**1.** What is antigenic drift and why is it important?

_____

_____

_____

_____

**2.** From your results, what is the identity of the hemagglutinin of the virus infecting the patient? How do you know?

_____

_____

_____

_____

**3.** What is the titer indicated by your results? How do you know?

_____

_____

_____

_____

**4.** Row C is serving as the control. What did you observe in row C, wells 1:10–1:1280? Explain this phenomenon. Would you consider this the positive or negative control?

_____

_____

_____

_____

**5.** What did you observe in row C, wells 1:2560–1:20480? Explain this phenomenon. Would you consider this the positive or negative control?

_____

_____

_____

_____

**6.** Why is it important to incubate the antibody with the sample/virus before combining this mixture with blood cells?

_____

_____

_____

_____

**7.** In 2013, there was an influenza outbreak in China thought to be the result of the evolution of a new strain caused by recombination of genes and then a species jump to humans. The serotypes involved were H7N3 isolated from domestic ducks, H7N9 isolated from wild birds, and H9N2 isolated from domestic poultry. What are the new viruses that could arise from recombination among these three viruses?

_____

_____

_____

_____

**8.** What term describes the jump of a virus between species?

_____

_____

_____

_____

# EnteroPluri-Test

## CASE FILE

Patient X developed a massive infection of a surgical wound after a routine surgical procedure. Four other patients, who also had surgery in the same unit as Patient X, were also found to be infected. DNA fingerprinting (exercise 27) revealed that one of the operating room nurses was the source of the infection.

©Gail Obenauf

The next step in this process is to identify the etiologic agent (organism) causing these infections. A culture was obtained from Patient X. A Gram stain of the wound culture revealed the presence of a gram-negative rod. You will use an *EnteroPluri-Test* to positively identify the organism causing the nosocomial infections in these patients. In addition to identifying the organism, it is also imperative that you determine the best course of treatment for Patient X. The Kirby-Bauer method (exercise 25) will be used to determine the antibiotic(s) most suitable for treatment of this infection.

## LEARNING OUTCOMES

At the completion of this exercise, students should be able to

- understand the basic function of multitest systems.
- demonstrate how to inoculate and read an EnteroPluri-Test.
- calculate an ID code from the results and use it to identify a bacterium.

## Background

In clinical laboratories, time, space, and expense are all important factors. They do not usually use individual media as you do in your educational lab. Instead, in order to identify bacteria from patient samples, many tests are done at the same time in one place. This involves using a **multitest system,** a single-piece unit that may run from 10 to over 100 tests at the same time. A number of different multitest systems are available for use in clinical labs, including the API, Minitek, PathoTec, and the one you will use today, the EnteroPluri-Test (formerly known as the Enterotube II).

Each of the 12 compartments in the EnteroPluri-Test contains a different kind of medium and may run one or several biochemical tests (there are 15 total). Several of the compartments are covered with a layer of wax to generate anaerobic conditions. When doing this procedure, focus on the overall function of the EnteroPluri-Test and how to generate the identification rather than focus on the reaction in each individual compartment. We will do many of the tests in the EnteroPluri-Test individually later in the term (such as the glucose fermentation, H₂S, urease, and citrate tests).

## ✅ Tips for Success

- Look carefully for the groove in the wire to break it–otherwise, it may stay bent instead of breaking.
- The reaction missed most frequently by students seems to be the gas reaction in the first compartment. Look carefully.

## Organisms (Plate Cultures)

- Patient X culture for diagnosis

## Materials (per Team of Two or Four)

- 1 EnteroPluri-Test unit
- Sharps container
- Indole reagent (period 2)
- EnteroPluri-Test code book or PDF file of codes (period 2)

## Procedure

### Period 1

1. Remove the blue cap. The wire at this end should be bent. This is the handle end of the EnteroPluri-Test. You will hold the wire at this end.
2. Remove the white cap. The wire at this end should be straight. This is the needle end of the EnteroPluri-Test. Do not touch this end of the wire–it is sterile (don't flame it, either).
3. Hold the EnteroPluri-Test almost parallel to the surface of the plate–angle it slightly. Using the *needle* end of the EnteroPluri-Test, scrape some bacteria from the surface of the plate.
4. Inoculate each compartment of the EnteroPluri-Test by slowly pulling and twisting the needle end through each compartment until you reach the last one. Do not completely remove the needle **(figure 36.1).**
5. Once you reach the last panel, slowly twist and push the needle all the way back into the EnteroPluri-Test.
6. Slowly pull the wire back out again, watching it carefully until you see a groove all the way around the wire. Once the groove is at the opening, bend the wire back and forth until the wire breaks at the groove.
7. Screw the caps back on each side of the EnteroPluri-Test.
8. You will then use the broken wire **(figure 36.2a)** to poke holes in the plastic over the slots **(figure 36.2b)** in eight of the compartments (adonitol through citrate) on the side. The other compartments do not have slots. Do not punch holes in the top.
9. Dispose of the wire piece in the sharps container.
10. Label the EnteroPluri-Test–you can write on the label where it says "NAME:" (figure 36.1)–and put it in the incubator.

### Period 2

1. Going one compartment at a time, compare your EnteroPluri-Test to the "uninoculated" (negative) and "inoculated" (positive) images in

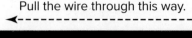
Pull the wire through this way.

**Figure 36.1** EnteroPluri-Test unit.
©Steven D. Obenauf

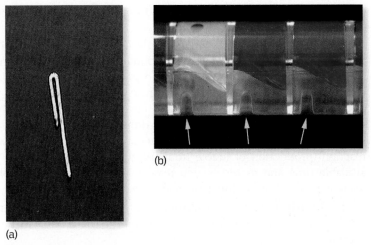

(a)

(b)

**Figure 36.2** **(a)** Wire piece for punching holes; **(b)** locations for punching holes. (a-b) ©Steven D. Obenauf

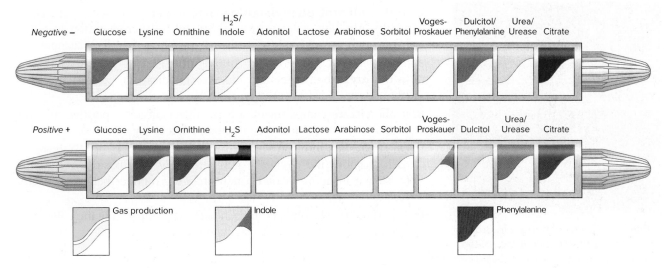

**Figure 36.3** Colors of positive and negative results for each chamber.
Courtesy and ©Becton, Dickinson and Company.

**figures 36.3, 36.4,** and **36.5** to determine which compartments are positive and which are negative. The reactions in each compartment and their colors are also listed in the Results and Interpretation section.

2. Use the results to generate the five-digit code.

3. Look that number up in the identification database of the EnteroPluri-Test code book or PDF file of codes to determine the organism's identity.

4. Record the name (or names if the code results in more than one) of the organism identified.

## Results and Interpretation

1. In the **glucose** compartment, you will read two results. A yellow color is positive for glucose fermentation; small bubbles in the agar or a gap between the agar and the wax is a positive result for gas production (figure 36.4).

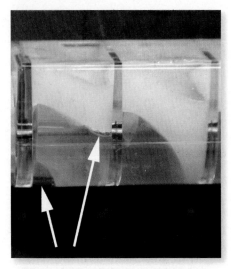

**Figure 36.4** Gas in the glucose compartment. ©Steven D. Obenauf

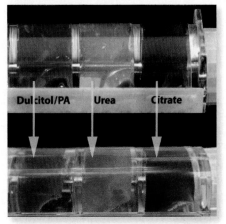

**Figure 36.5** Color changes in Group 5 compartments. Uninoculated on top, inoculated below. ©Steven D. Obenauf

2. In the **lysine** and **ornithine** compartments, a purple color indicates a positive reaction (decarboxylation).

3. In the **hydrogen sulfide (H₂S)/indole** compartment, you will read two results. A black color forming around the wire in the center of the compartment is positive for the production of H₂S. To do the indole test, read all of the other reactions first. Then place the EnteroPluri-Test with the thin plastic covering facing up. Punch a small hole in the covering and add a drop of indole reagent to the chamber. If the reagent turns bright red, this indicates a positive reaction (production of indole from tryptophan). Depending on available time and materials, you may or may not add the indole reagent to the EnteroPluri-Test. If not, record it as negative.

4. In the **adonitol, lactose, arabinose,** and **sorbitol** compartments, a yellow color indicates a positive reaction (fermentation of these sugars).

5. The Voges-Proskauer compartment is used to confirm certain identifications. It is not part of generating the ID code, so you will not read this result.

6. In the **dulcitol/phenylalanine** compartment, you will read two results. A yellow color is positive for dulcitol (fermentation of this sugar). A black or smoky gray color is positive for phenylalanine (deamination).

7. In the **urea** compartment, a pink color indicates a positive reaction (the presence of urease) (figure 36.5).

8. In the **citrate** compartment, a blue color indicates a positive reaction (the utilization of citrate as a carbon source) (figure 36.5).

9. On the results form on your lab report, record a "1" or "2" above each chamber and circle the number below each positive reaction (as shown in figure 36.6). Do not circle the number below negative reactions.

10. Add the circled numbers in each section/group to obtain a five-digit number. Write it in the boxes below each section.

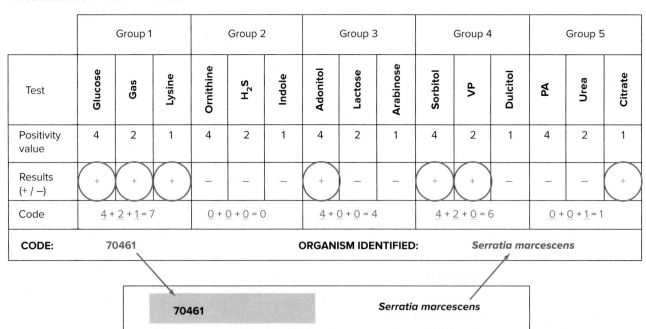

| Test | Group 1 | | | Group 2 | | | Group 3 | | | Group 4 | | | Group 5 | | |
|---|---|---|---|---|---|---|---|---|---|---|---|---|---|---|---|
| | Glucose | Gas | Lysine | Ornithine | H₂S | Indole | Adonitol | Lactose | Arabinose | Sorbitol | VP | Dulcitol | PA | Urea | Citrate |
| Positivity value | 4 | 2 | 1 | 4 | 2 | 1 | 4 | 2 | 1 | 4 | 2 | 1 | 4 | 2 | 1 |
| Results (+ / −) | + | + | + | − | − | − | + | − | − | + | + | − | − | − | + |
| Code | 4 + 2 + 1 = 7 | | | 0 + 0 + 0 = 0 | | | 4 + 0 + 0 = 4 | | | 4 + 2 + 0 = 6 | | | 0 + 0 + 1 = 1 | | |

**CODE:** 70461     **ORGANISM IDENTIFIED:** *Serratia marcescens*

70461     *Serratia marcescens*

**Figure 36.6** Calculating the ID code.
Adapted from Becton, Dickinson and Company.

Name _____

Date _____

# EnteroPluri-Test

## Your Results and Observations

Record your results and look up the code they generate in the EnteroPluri-Test code book or PDF file of codes. Variations in test results lead to one organism having more than one code. For example, possible codes for *Klebsiella pneumoniae* include 70771 and 40753, *E. coli* includes 75340 or 41150, and *Proteus* may be 42007 or 43006.

| Test | Group 1 | | | Group 2 | | | Group 3 | | | Group 4 | | | Group 5 | | |
|---|---|---|---|---|---|---|---|---|---|---|---|---|---|---|---|
| | Glucose | Gas | Lysine | Ornithine | $H_2S$ | Indole | Adonitol | Lactose | Arabinose | Sorbitol | VP | Dulcitol | PA | Urea | Citrate |
| Positivity value | 4 | 2 | 1 | 4 | 2 | 1 | 4 | 2 | 1 | 4 | 2 | 1 | 4 | 2 | 1 |
| Results (+ / −) | | | | | | | | | | | | | | | |
| Code | _ + _ + _ = | | | _ + _ + _ = | | | _ + _ + _ = | | | _ + _ + _ = | | | _ + _ + _ = | | |
| **CODE:** | | | | | | | **ORGANISM IDENTIFIED:** | | | | | | | | |

Adapted from Becton, Dickinson and Company.

## Interpretation and Questions

1. What organism is Patient X infected with?

_____

_____

_____

_____

# NOTES

EXERCISE

# 37

# Soil Microbiology

## CASE FILE

A 40-year-old wildlife biologist working in the western United States went to a local clinic complaining of vomiting, diarrhea, fever, and abdominal pain. The symptoms had begun 24 hours earlier. The patient was given a prescription for an analgesic and an antiemetic and released. The patient returned the next day, reporting worsening symptoms. Prominent axillary bubos were noted and bubonic plague was suspected. The  Source: Centers for Disease Control
patient was admitted to the hospital and treated with gentamicin and doxycycline. Blood cultures were positive for *Yersinia pestis*. The patient succumbed to the infection 2 days later.

The patient had reported recently performing a necropsy on a mountain lion that died of indeterminate causes. Follow-up studies revealed that the mountain lion was seropositive for exposure to *Y. pestis*. Plague can be contracted by contact with contaminated meat or fur and from bites from infected fleas. Rodents are the natural reservoir for the disease, although survival of the organism in soils has been documented. *Y. pestis* is thought to have been introduced into the United States by rats traveling on ships from China. It is found predominantly in the western United States. Plague can be successfully treated when caught early. However, the disease progresses rapidly, with the onset of symptoms within 2 to 6 days of exposure. Untreated, mortality from this disease is 40% to 60%.

## LEARNING OUTCOMES

At the completion of this exercise, students should be able to
- describe how to perform a serial dilution.
- calculate the colony-forming units of a sample, given the dilution and colony count.
- identify yeasts, molds, bacteria, and actinomycetes when growing on an agar plate.
- explain the purpose of using glycerol yeast extract (GYE), Sabouraud agar (SAB), and nutrient agar (NA) in this exercise.

## Background

Soil is alive! While that may sound like a trite line from a science fiction movie, it is actually true. Millions and, in some cases, billions of organisms can be found in a gram of soil. These organisms include representatives from every domain or kingdom and the viruses. Bacterial cells alone contribute nearly a ton of mass to a single acre of soil. There are many organisms in soil that to date have not been isolated. More than 4,000 species of prokaryotes alone have been identified in a single gram of soil. For purposes of this exercise, only bacteria and fungi will be discussed.

Soil varies significantly from place to place. Soils are often characterized by the amount of organic material present, the mineral composition, and particle size. The amount of organic material, in particular, is important to microbial communities. Because of the complexity of soil structure and composition, soil microbial content can vary dramatically within a few meters. Some areas are very moist and nutrient rich, while others are drier and more nutrient poor. Soils around plant roots (the zone called the rhizosphere) are rich with plant exudates and are, therefore, particularly rich in microbiota. Each particle of soil represents its own microenvironment. Each grain in a soil represents a microbial universe.

## Soil Bacteria

Soil bacteria serve several functions in soil communities. Most soil bacteria are decomposers. They break down and recycle organic material in the soil and improve soil agronomic productivity. These organisms are critically important to the cycling of carbon compounds, like grass clippings, leaf litter, and dead animals. They help prevent nutrients from leaching out of the soil. Some bacteria form mutualistic relationships (a type of symbiosis where both partners benefit) with plants. Members of the genus *Rhizobium* live within plant roots. These bacteria have the capability to fix nitrogen—that is, they convert atmospheric nitrogen ($N_2$) to an organic form of nitrogen. The bacteria provide the plant with nitrogen, a limiting element in most soils, and the plant provides the bacteria with a place to live and nourishment. Some bacteria are plant pathogens. The final bacterial category includes the chemolithotrophic bacteria. These organisms are important in phosphorus and nitrogen cycling. Soils and nitrogen cycling are discussed in another exercise.

The prokaryotic actinomycetes are members of the soil community that are medically very important. Some actinomycetes produce antibiotics. Members of the genus *Streptomyces* and related organisms produce antimicrobials, including streptomycin, tetracycline, vancomycin, nystatin, and ivermectin. Actinomycetes give soil its characteristic earthy scent. Actinomycetes as a group are very unusual. They exhibit a filamentous growth form that under the microscope looks like very fine fungal filaments. Actinomycetes also exhibit other novel forms of metabolism and can break down recalcitrant compounds, like the complex carbohydrates cellulose and chitin.

## Soil Fungi

Fungi are also very numerous in soil communities. In fact, the largest organism on earth today is a fungus that spreads over 2,200 acres in Oregon and may be more than 2,400 years old. Genetic testing from various locations miles apart indicate that this is one organism.

Fungi, like bacteria, fill different niches within the soil community. Fungi are decomposers. They break down complex organic molecules and other compounds and release the components of these materials back into the soil. This process enriches the soil. Fungal filaments modify soil structure by binding soil particles together. By modifying soil structure, fungi can improve water and nutrient retention. Fungi also can function as parasites and pathogens. Soil fungi are major pathogens of plants, and they have a significant impact on plant productivity and agricultural profits. However, some fungi also serve as biocontrols for other plant pathogens, like the nematodes. Nematodes are roundworms. Some nematodes attack plant roots and dramatically affect plant vigor and may kill the plant. There are fungi that "hunt" nematodes. They use their hyphae to form little nooses. When a nematode wiggles into the noose, the hypha traps and digests the nematode. Some fungi participate in mutualistic symbiotic associations with plants called mycorrhizae. The fungal network increases the flow of water and nutrients back to the plant root. These associations are particularly important to the plant, since fungi very effectively scavenge and transport phosphorus back to the plant. The plant in turn provides nutrients to the fungus. These associations improve plant productivity. Mycorrhizal associations exist with most trees and many agricultural crops. In fact, many seeds are soaked in mycorrhizal dips to inoculate them before planting.

## Counting Soil Microbes

Scientists have determined that to accurately count plated bacteria, there should be between 30 and 300 colony-forming units per Petri dish. Because soil microbe numbers are so high, the samples must be serially diluted to achieve accurate counts of the microbiota. Aliquots of the diluted sample are then mixed with molten tempered agar to produce a **pour plate.** The molten agar will harden at room temperature and "lock" the organisms in place. Organisms will grow on the surface of the agar and throughout the agar depending on their oxygen requirements.

Selective media are used to encourage the growth of specific microbe populations. In this exercise, a general growth medium will be used in addition to media that encourage the growth of fungi and actinomycetes. Nutrient agar (NA) will be used to quantitate the total number of colony-forming units per gram of soil. Nutrient agar is a general nonselective, nondifferential growth medium with a neutral pH that will allow the growth of many bacteria and fungi. Sabouraud (SAB) agar will be used to quantitate the mold and yeast population. Sabouraud agar is a medium used to encourage the growth of fungi. The pH of the medium is 4.5. The acidity of this medium inhibits bacterial growth but allows fungi to grow. This medium may have antibiotics added to it to further inhibit bacterial growth. Yeast are fungi that grow as single cells or pseudohyphae. Yeast colonies will have a slimy, mucoid appearance and look very much like bacterial colonies. Molds are filamentous fungi and have a fuzzy appearance. Glycerol yeast extract (GYE) agar is the final medium to be used in this exercise. GYE agar is used for the isolation of prokaryotic actinomycetes. Glycerol, a sugar alcohol, is the sole source of carbon in this medium and is not easy for most organisms to metabolize. Actinomycetes, with their novel metabolic pathways, can easily metabolize glycerol. In addition, this medium has an alkaline pH. The alkaline pH favors the growth of the actinomycetes while inhibiting the growth of most bacteria and fungi. Actinomycete colonies have a ground-glass to powdered sugar appearance. Many of them are colored.

## ✔ Tips for Success

- Shake the soil-containing dilution bottles for a least 1 minute.
- Remove and pour melted and tempering agar deeps one at a time. The agar solidifies quickly. If you leave it on your desk, it will harden before you can use it.
- Make sure you are dispensing the sample into the bottom of the Petri dish, not the Petri dish lid.
- Uncover the plates only to add the sample and agar. Do not remove the plate lids and leave them on the table. Your samples will be contaminated by organisms from the room.
- Mix sample with the agar by gently swirling the plate with a figure-eight motion. **Gently!**

- Do not move the plates for at least 5 minutes after pouring the agar deeps into the plates. This allows the agar in the plates to cool and solidify. Invert the plates and place them in the incubator (28°C).
- Most of the colonies growing on the GYE plates will probably **not** be actinomycetes. You must identify and count them based on the colony morphology.

## Materials

- Soil
- Balance
- Weigh boat
- Scoopula
- 2 dilution bottles (French squares) containing 99 mL of peptone broth
- 1-mL pipettes
- Pi-pump

- Micropipettor (0 to 200 µL)
- Micropipettor tips
- 9 empty sterile Petri dishes
- 3 Sabouraud agar deeps
- 3 nutrient agar deeps
- 3 glycerol agar deeps
- Hot water bath
- Bunsen burner

## Procedure

### Period 1

1. Turn on the balance. Make sure the balance is reading in grams; a "g" should be visible on the right side of the balance display area.
2. Gently place the weigh boat on the balance pan. Press the tare button. This will zero out the balance.
3. Use the scoopula to transfer 1 gram of soil into the weigh boat.
4. Transfer the soil to a dilution bottle by gently folding the weigh boat and tipping it into the bottle.
5. Recap the dilution bottle and shake vigorously for 1 minute. Bacteria adhere tightly to the soil particles. Shaking helps dislodge organisms from the soil particles.
6. Use the 1-mL pipette and pi-pump to transfer 1 mL of solution from this bottle into the other 99-mL peptone dilution bottle.
7. Shake the bottle vigorously for 1 minute.
8. Get nine empty Petri dishes from the supply table. Label plates as shown in **figure 37.1**. Indicate the agar type and the dilution.
9. Borrow a micropipettor from the supply table. Set the micropipettor to dispense 100 µL. Your instructor can assist you if you are unfamiliar with the pipettor. Attach pipettor tip.
10. Use the pipettor to transfer a 100 µL sample from the second dilution bottle ($10^{-4}$) to the plate labeled "NA $10^{-5}$." Repeat this step for the plate labeled "GYE $10^{-5}$" and the plate labeled "SAB $10^{-5}$."
11. Using the same pipette tip to transfer a 1-mL sample from the first dilution bottle ($10^{-2}$) to the plate labeled "NA $10^{-2}$." Repeat this step for the plate labeled "GYE $10^{-2}$" and "SAB $10^{-2}$."
12. Discard the tip in the biohazard bag and return the pipettor to the common materials bench.
13. Acquire a 1-mL pipette and pi-pump. Remember, the pipette is sterile until you remove it from the package. Handle the pipette carefully; do not touch the tip.
14. Use the pipette to transfer a 1-mL sample from the second dilution bottle ($10^{-4}$) to the plate labeled "NA $10^{-4}$." Repeat this step for the plates labeled "GYE $10^{-4}$" and "SAB $10^{-4}$."
15. Remove one NA deep from the water bath. Use a paper towel to dry off the tube.

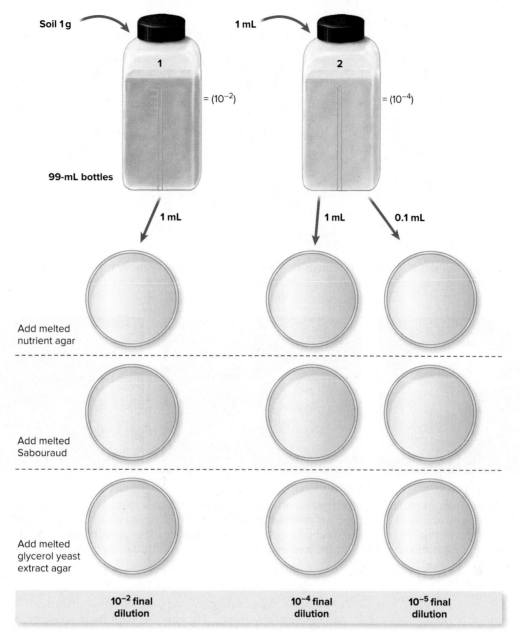

**Figure 37.1** Dilution scheme.

16. Remove the cap and flame the mouth of the tube. Carefully lift the Petri dish lid and aseptically dispense the agar into one of the "NA" labeled Petri dishes. Swirl the plate gently, as shown in **figure 37.2.**

17. Repeat steps 15 and 16 for both the two remaining plates labeled with "NA."

18. Repeat steps 15, 16, and 17 for the plates labeled with "GYE" and "SAB."

19. Discard empty agar deeps in the discard area.

20. Discard pipettes appropriately.

21. Place full dilution bottles in the discard area. DO NOT dump the water and soil down the drain. Do not place the dilution bottles in the biohazard bags.

## Period 2

1. Examine your plates for growth. Distinguish among bacterial, fungal (yeast, mold), and actinomycete colonies.

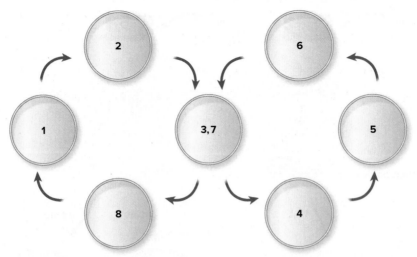

**Figure 37.2 Rotating the Petri dish to mix inoculum with agar.** Gently and slowly move the Petri dish in a figure-eight pattern. This will mix the agar and the sample together. Be careful not to splash agar on the lid.

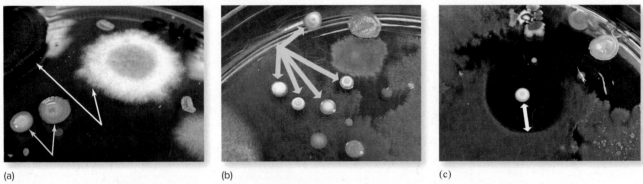

(a)  (b)  (c)

**Figure 37.3 Soil microbes.** In panel **(a)**, yeasts (yellow arrows) and molds (white arrows) are shown growing on Sabouraud agar. Some examples of actinomycetes colonies growing on glycerol yeast extract agar are shown at the yellow arrow tips in panel **(b)**. The colony in panel **(c)** is producing an antibiotic. The white, double-headed arrow indicates the zone of inhibition. An antibiotic produced by this colony diffused through the agar and prevented the growth of other bacteria. (a-c) ©Susan F. Finazzo

2. Many soil organisms grow very slowly. Your instructor may suggest that you allow these plates to incubate for a week or more before you record your results.

3. Determine which plates are countable. Calculate the CFU for your countable plates.

4. Your instructor may have you stain colonies of interest.

5. Dispose of your plates as instructed.

## Results and Interpretation

Soils support a diverse microbiota whose members exhibit interesting and unique biosynthetic pathways. Many of our current chemotherapeutic agents were originally isolated or derived from soil-dwelling organisms. Isolation of particular groups of organisms from soils requires specialized agar. Sabouraud agar enhances the growth of fungi. Yeasts growing on Sabouraud agar **(figure 37.3a)** have a colonial form and appear shiny and moist. Molds will appear fuzzy. Glycerol yeast extract agar can be used to isolate actinomycetes. Actinomycetes **(figure 37.3b)** form discrete colonies with a crystalline appearance. Many members of this group are known to produce antibiotics **(figure 37.3c)**.

Name _____

Date _____

# Soil Microbiology

| Agar type | Microbe type | Dilution of plate counted | Count | Quantity: CFU (cells)/gram |
|-----------|--------------|---------------------------|-------|----------------------------|
| Nutrient agar | Bacteria | | | |
| Glycerol yeast extract agar | Actinomycetes | | | |
| Sabouraud agar | Fungi: molds | | | |
| | Fungi: yeasts | | | |

## Interpretation and Questions

**1.** Which type of microorganism was the most abundant in the soil?

_____

_____

_____

_____

**2.** Some actinomycetes produce substances (antibiotics) to kill off other microbes around them. Do you see any evidence of this?

_____

_____

_____

_____

**3.** Why was it necessary to use three different types of agar?

_____

_____

_____

**4.** Why are pour plates, not spread plates, used in this exercise?

_____

_____

_____

_____

**5.** Many soil microbes have been impossible to isolate. How do you explain this? What would you suggest to improve isolation of these "unculturable" organisms?

_____

_____

_____

_____

# Hydrolytic Enzymes and Disease Mechanisms

# DNase Test

## CASE FILE

A 66-year-old male was seen in the emergency room complaining of fever, chills, and difficulty breathing. The patient reported that he had started to feel ill 24 hours earlier. The patient presented with a temperature of 40°C (104°F). Auscultation of the lungs revealed bilobar congestion. The patient's history revealed that he was in stage 5, or end-stage, renal disease and had been undergoing kidney dialysis via an arteriovenous fistula for the past 4 years. The patient was admitted to the hospital. The doctor ordered a chest X ray and a blood culture. The chest X ray revealed multilobe consolidation. The blood culture indicated the presence of coagulase-positive, DNase-positive bacteria. A diagnosis was made of *Staphylococcus aureus* sepsis with secondary staphylococcal pneumonia.

©Image Source/Getty Images RF

Antibiotic-sensitivity testing (as you performed in exercise 25) indicated the strain was methicillin resistant. The patient was treated with the antibiotic vancomycin. Initially, the infection appeared to respond to treatment; however, the patient relapsed 1 week after the initiation of treatment and subsequently died of multiple organ failure.

Central line–associated bloodstream infections (CLABSIs) number in the tens of thousands each year. *S. aureus,* a normal inhabitant of the skin and nasal passageways, is one of the leading causative agents of these infections. Dialysis patients have nearly a 100-fold increased risk of these infections when compared to the general population. Health care workers play a key role in avoiding these infections. Prevention may involve strictly following hygienic measures and cleaning the exit site with antiseptics. The use of topical antimicrobials at the catheter site has proven to be particularly effective. The use of medical grade honey has even been attempted (with limited success).

## LEARNING OUTCOMES

At the completion of this exercise, students should be able to
- define *extracellular enzyme/extracellular digestion.*
- recall the clinical importance of the DNase test.
- perform a DNase test.
- explain the appearance of a positive and negative test result.

## Background

DNase is an extracellular enzyme–that is, it is produced within the cell and secreted outside of the cell wall. Casease (caseinase), gelatinase, and amylase–discussed in this section of the laboratory manual on hydrolytic enzymes–are also extracellular enzymes. Once secreted, extracellular enzymes diffuse into the environment and act upon their substrate. In the case of DNase, the enzyme breaks down macromolecular DNA to the level of the mono- or polynucleotide. Nucleotides are the building blocks of nucleic acids. These nucleotide subunits could serve as nutrients for the bacteria. In addition, the production and secretion of DNase have been linked to an organism's ability to degrade biofilms, which would enhance the organism's ability to invade the host or spread from the site of origin. This test is clinically important in the identification of pathogenic *Staphylococcus aureus*.

DNase agar is made by mixing powdered DNA into nutrient agar. In some instances, dyes such as methyl green or toluidine blue are added to the medium. The DNase agar in this exercise does not contain dye and looks very much like nutrient agar. All of the organisms used in this exercise will grow on this medium. The presence of DNA is not visible in the medium until the reagent (hydrochloric acid [HCl]) is added. The test is done by first inoculating and incubating a DNase plate. After incubation, HCl is poured onto the surface of the agar plate. Hydrochloric acid is a potent denaturant. As DNA in the agar denatures, its structure changes and the agar will turn white wherever intact DNA is still present in the medium (refer to figure 38.2). If DNase has been secreted into the medium, the DNA in the agar surrounding the colony will be degraded and the medium will not change color. A positive test result for the production of DNase is a clear zone around the colony with the addition of the reagent, hydrochloric acid. A negative result (no DNase produced) is a whitening of the agar surrounding the colony.

###  Tips for Success

- Inoculate and incubate the DNase agar plate first. HCl is added during the next lab period.
- Do not make the streaks too long, or the zone from one streak can "bleed" over onto the other.
- **Be careful** moving the plate once the HCl has been added. HCl is an eye and respiratory irritant.
- Placing the plate against a dark background will make it easier to observe the zone of clearing.

### Organisms (Slants or Broth Cultures)

- *Staphylococcus aureus*
- *Enterobacter aerogenes*

### Materials (per Team of Two or Four)

- 1 DNase plate
- Dropper bottle of 1N HCl (period 2)
- Loop
- Bunsen burner
- Microincinerator

### Procedure

#### Period 1

1. Obtain a DNase agar plate and label it with your group name and date.
2. Use your indelible marker to draw a line on the bottom of the Petri dish to divide the plate in half **(figure 38.1)**.

**Figure 38.1** Streak pattern of DNase plate inoculation.

3. Label each half of the plate with the organism's name as previously listed.

4. Streak inoculate each organism onto one-half of the plate.

5. Place the plate in the incubator agar side up.

## Period 2

1. Remove your plate from the incubator.

2. Remove the Petri dish lid and add HCl until the entire plate is covered with acid. Replace the lid.

3. Wait 10 minutes and then observe and record your results.

4. Carefully discard the Petri dish in the biohazard bag.

## Results and Interpretation

DNase medium is made by adding DNA to a nutrient agar. As a general growth medium, most of the commonly used organisms will grow on these plates. Some organisms produce and secrete an extracellular DNase that breaks down DNA in the surrounding environment. The smaller breakdown products diffuse back to the cell and are absorbed as nutrients. The presence (or absence) of DNA in the medium is detected using the reagent, HCl. After incubation, HCl is poured onto the plate, denaturing DNA molecules in the medium. If an organism produces DNase **(figure 38.2),** then the medium surrounding the colony is free of DNA and remains clear (DNase-positive). If the organism does not produce DNase, then the DNA in the medium surrounding the colony will denature and turn the medium opaque (white).

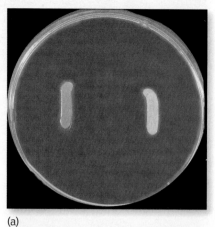
(a)

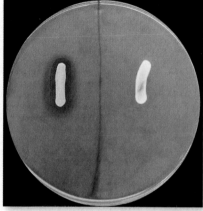
(b)

**Figure 38.2 DNase plate.** The image on top is the incubated plate before the addition of the reagent, HCl. HCl was poured onto the plate to produce the result observed in the bottom panel. Which side of the plate, right or left, indicates a positive DNase test? On which side of the plate, right or left, was DNase secreted? (a-b) ©Susan F. Finazzo

# NOTES

# DNase Test

## Your Results and Observations

Draw your plate and/or photograph with your mobile device before and after the addition of HCl.

**Before**

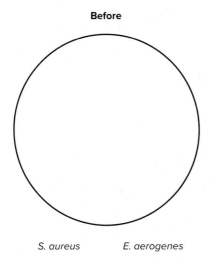

S. aureus          E. aerogenes

**After**

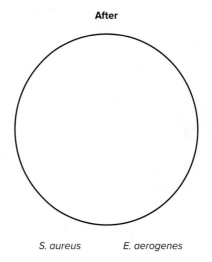

S. aureus          E. aerogenes

## Interpretation and Questions

**1.** Which organism secreted DNase?

_____

_____

_____

_____

_____

**2.** Why would an organism's ability to produce this enzyme be clinically significant?

_____

_____

_____

_____

_____

**3.** *S. aureus* is a common component of the biota of the skin and nasal passages. What steps should be taken by the dialysis unit described in the case file to minimize nosocomial infections in its patients?

_____

_____

_____

_____

_____

# Gelatin Hydrolysis

## CASE FILE

A 63-year-old man was seen in the emergency room. He had
been golfing earlier in the day and had scratched his lower
leg on some brush while retrieving a ball from the rough.
A large area surrounding the cut exhibited edema and ery-
thema. The inflammatory response was diffuse and spreading
and encompassed the entire lower appendage. The tissue
immediately surrounding the abrasion was shiny and taut. The

©McGraw-Hill Education/
Don Rubbelke, photographer

patient appeared pale and had a blood pressure of 105/60
mmHg, a pulse of 110 beats per minute, and a temperature of 40°C (104°F).
The patient complained of extreme pain in his lower leg. The doctor
admitted the man to the hospital because of the patient's symptoms
(low blood pressure, tachycardia, fever, and extreme sensation of
pain) and his preexisting chronic health condition (diabetes mellitus).
The doctor suspected necrotizing fasciitis (NF). He requested blood
work, radiologic testing, and histological testing be done to confirm the
diagnosis. Additionally, he requested immediate surgical debridement
(removal of dead or diseased tissue) beyond the margin of the wound
and the immediate administration of intravenous broad-spectrum
antibiotics.

The patient's blood work was consistent with NF infection. A
computed tomography (CT) scan revealed gas pockets within the
tissue. During surgical debridement, little bleeding was evident and
dishwater pus exuded from the surrounding tissue, indicating liquefac-
tion of the tissue. Gram staining of biopsied tissue from the wound area
indicated the presence of a gram-negative rod. Further testing identified the organism as *Serratia marcescens*. The organism was
collagenase- and gelatinase-positive.

The patient remained hospitalized for 2 weeks and underwent daily
wound debridement, concurrent antibiotic therapy, and supportive care.
NF is a rare but potentially fatal condition. Mortality rates can exceed
70%. There are a number of risk factors that predispose patients to NF,
including diabetes, arteriosclerosis, AIDS, and any condition that causes
immunosuppression. Mortality in NF patients increases dramatically with
age. Rapid diagnosis is essential for a successful outcome.

At the completion of this exercise, students should be able to

- explain the appearance of a positive and negative gelatin hydrolysis test result.
- explain the basis for the gelatin hydrolysis test.

## Background

Gelatin is a high-molecular-weight, water-soluble protein derived from collagen, a component of connective tissues of animals. It is translucent and tasteless. Gelatin is a long, fibrous molecule that when chilled on ice or in the refrigerator forms cross-links to itself to create a semisolid gel.

As a protein, gelatin could provide a rich source of amino acids and peptides to bacteria. Unfortunately, the protein is too large to be transported into the bacterial cell intact. It must first be broken down by extracellular proteases, like gelatinase. Gelatinase hydrolyzes bonds in the gelatin molecule to produce smaller peptides and amino acids **(figure 39.1).** These smaller components can diffuse more easily through the medium and then be transported into the bacterial cell.

Nutrient gelatin medium is used to detect the production of an extracellular gelatinase. The medium is produced by combining nutrient broth and gelatin. This medium contains no agar. At room temperature, the medium can be liquid to slightly viscous. When refrigerated, the gelatin molecules cross-link and form a semisolid gel. The medium is inoculated with the test organism and incubated for 24 to 48 hours (or longer). After incubation, the tubes are placed in the refrigerator or on ice for 20 to 30 minutes. If the tube gels (forms a semisolid gel) after incubation and refrigeration, then the organism is negative for the enzyme gelatinase **(figure 39.2).** The enzyme is not present; therefore, the tube gels as expected. If, after incubation and refrigeration, the inoculated tube is liquid (does not resolidify) when tilted, then the test is positive and the organism produced the enzyme gelatinase. The enzyme was secreted into

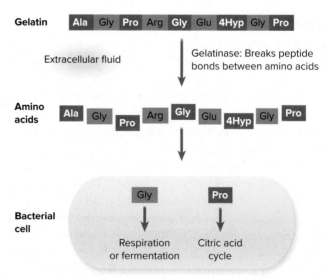

**Figure 39.1 Gelatinase activity.** Gelatinase is an extracellular protease that hydrolyzes gelatin to produce short peptides and amino acids. These smaller components diffuse more easily and can be transported into the cell.

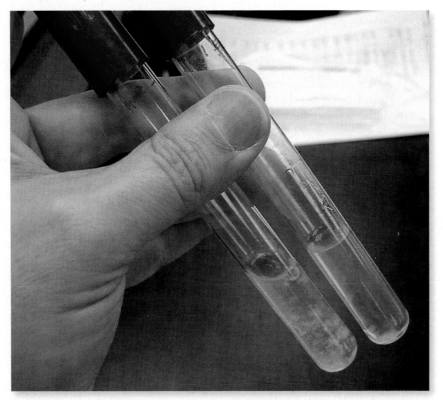

**Figure 39.2 Gelatinase activity.** These tubes were inoculated, incubated for 48 hours, and then placed in an ice bath for 20 minutes. The tube to the left is runny (liquid), indicating that gelatin molecules did not cross-link and gel. This is a positive test result. The tube to the right gelled, indicating gelatinase was not produced and that gelatin was not hydrolyzed. This is a negative test result. ©Susan F. Finazzo

the medium and hydrolyzed the gelatin to produce short peptides and amino acids. These shorter strands cannot cross-link effectively to produce the gel, and therefore the medium remains liquid.

## ✓ Tips for Success

- After inoculation, place the tubes in the incubator, not the refrigerator.
- After incubation, place the tubes on ice or in the refrigerator for at least 20 minutes.
- The expression of gelatinase activity can require up to a week of incubation. Your instructor may ask that you reincubate any negative tubes to ensure that bacteria have had an adequate time to grow.
- Be careful when tipping the tubes to not spill liquefied gelatin on yourself or your work surface.

## Organisms (Slant or Plate Cultures)

- *Staphylococcus aureus*
- *Staphylococcus epidermidis*

## Materials (per Team of Two or Four)

- Inoculating needle or loop
- 2 gelatinase medium tubes
- Bunsen burner
- Microincinerator

## Procedure

### Period 1

1. Label your tubes with the organism's name.

2. Use your inoculating needle to aseptically transfer the appropriate organism to each tube. Stab to the bottom of each tube. If the medium has just been removed from the refrigerator, it may be semisolid and difficult to stab. If the medium is more liquid, you may use either the needle or the loop for inoculation. In either case, stab to the bottom of the tube.

3. Place the tubes in the incubator and incubate for 24 to 48 hours. An uninoculated tube should also be incubated as a negative control. One control tube per class is adequate.

### Period 2

1. Remove the tubes from the incubator and place the tubes in the refrigerator or in an ice bath for 20 to 30 minutes.

2. Gently tip the tubes. Do not completely invert the tubes.

3. Record your results. See figure 39.2 to interpret your results.

4. Your instructor may suggest reincubating negative test tubes until the next class.

## Results and Interpretation

Gelatin is a long, fibrous protein that cross-links to form a semisolid gel. Some bacteria produce a protease exoenzyme, gelatinase, that breaks down gelatin to the amino acid or small peptide level. After incubation, the medium is chilled in the refrigerator or in an ice bath. If the medium gels, then gelatin molecules remained intact and cross-linked normally (figure 39.2). This organism is gelatinase-negative. If, on the other hand, the medium does not gel and remains liquid, then the gelatin molecules were digested and could not cross-link. This organism is gelatinase-positive.

This medium is used to differentiate *Staphylococcus aureus* (gelatinase-positive) from *Staphylococcus epidermidis* (gelatinase-negative). This test is also useful for differentiating *Pseudomonas aeruginosa* (gelatinase-positive) from the other gram-negative bacteria. Gelatinase and collagenase play a role in the ability of some bacteria to cause disease. The ability to break down connective tissue increases the ability of the pathogen to spread through the body and cause tissue damage. If you do a Web image search for necrotizing fasciitis, you will see some disturbing images showing the destructive power of some bacteria because of the enzymes they make.

Name _____

Date _____

# Gelatin Hydrolysis

## Your Results and Observations

Record your results after incubation and refrigeration in the table.

| Organism | Results (solid or liquid) | Does this organism produce gelatinase? |
|---|---|---|
| *Staphylococcus epidermidis* | | |
| *Staphylococcus aureus* | | |

**1.** Why is this test typically done in a test tube and not in a Petri dish?

_____

_____

_____

_____

_____

**2.** Clinically, how would the ability to produce gelatinase or a similar protease be an advantage to an invasive bacterium?

_____

_____

_____

_____

_____

**3.** Why must the tubes be interpreted after incubation and **refrigeration?**

_____

_____

_____

_____

_____

**4.** You are testing an unknown organism from an infected patient. Your gelatin tube results are positive for gelatinase. Additional testing shows that the organism is negative for glucose and lactose fermentation, and negative for $H_2S$ production. Using the table in exercise 49 (49.2), what is the organism?

_____

_____

_____

_____

_____

# Starch Hydrolysis

## CASE FILE

Two premature infants in a neonatal intensive care unit (NICU) began to show temperature fluctuations, increased signs of respiratory distress, and below-normal pulse oximeter readings. Blood samples were sent to culture for bacteria, and tracheal aspirates were gram stained and sent for culture. No organisms were detected in the blood cultures, but the tracheal aspirates were positive for bacteria. Further testing showed the presence of *Bacillus cereus*. Testing of hospital linens, surfaces in the NICU, and ventilator equipment as well as interviews with respiratory therapists and nurses were undertaken. Cultures from the ventilator flow sensors were positive for *B. cereus*, indicating ventilator-associated pneumonia (VAP) as a likely cause of the infants' symptoms. Disinfection procedures used for the sensors were reviewed.

©SPL/Science Source RF

### LEARNING OUTCOMES

At the completion of this exercise, students should be able to

- explain the appearance of a positive and negative starch hydrolysis test result.
- explain the basis for the starch hydrolysis test.
- identify the substrate, enzyme, product, and reagent used in the starch hydrolysis test.

## Background

Starch is an abundant, naturally occurring plant polysaccharide. Starch is so abundant that it comprises 60% to 80% of the dry weight of a potato! What we generically call starch is actually a mixture of amylose and amylopectin. Amylose and amylopectin are both composed of repeating units of glucose joined by glycosidic bonds **(figure 40.1)**. Amylose and amylopectin simply differ in their branching pattern. Starch is a rich source of carbon and simple sugars.

Amylase is an extracellular enzyme that catalyzes the hydrolysis of amylose **(figure 40.2)** and, to a limited extent, amylopectin. The enzyme breaks the glycosidic bonds to produce maltose (disaccharide), glucose, and

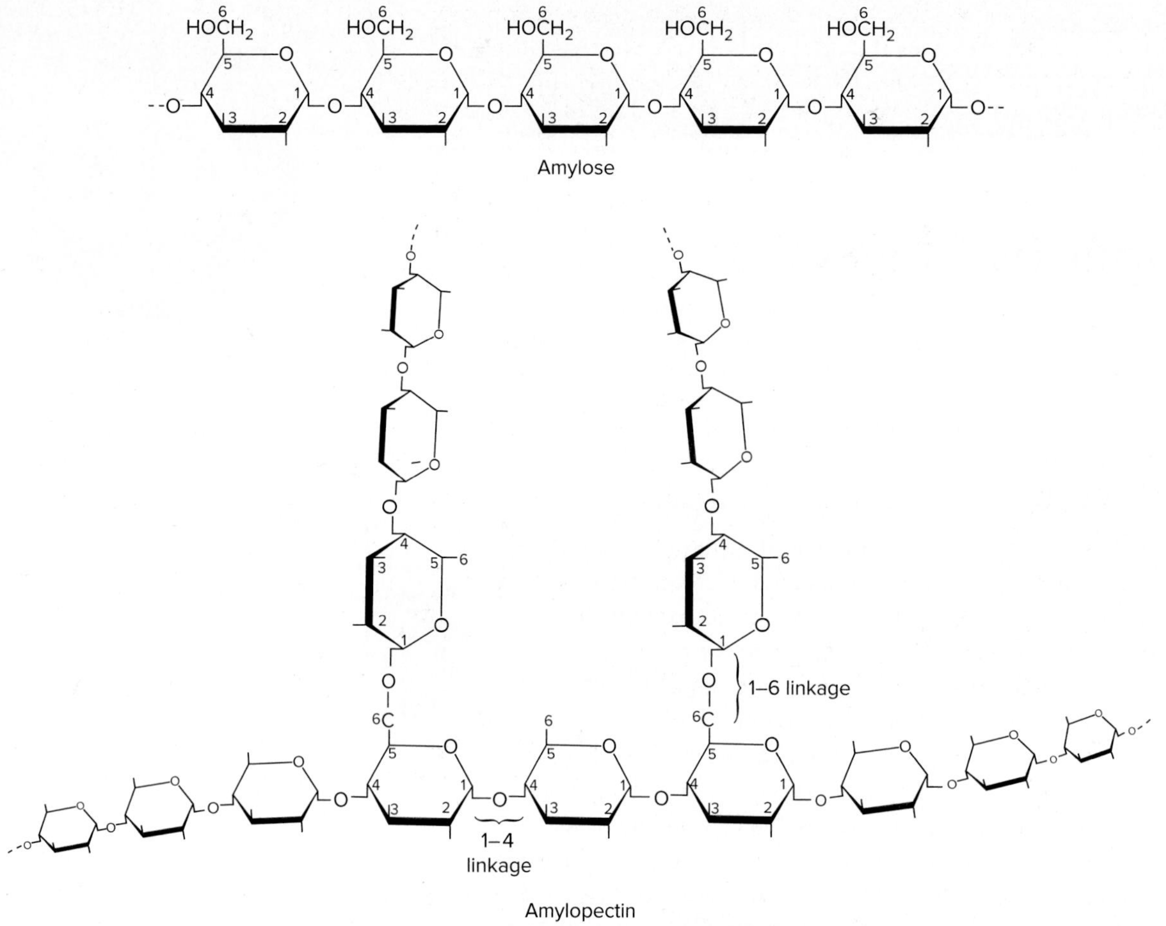

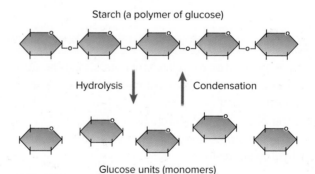

**HOCH₂** appears in the structure labels as **HOCH₂** across the amylose chain.

Amylose

1–6 linkage

1–4 linkage

Amylopectin

**Figure 40.1 Structure of amylose and amylopectin.** Amylose and amylopectin are two polysaccharides that are found in starch. They are composed of repeating units of the monosaccharide glucose.

Starch (a polymer of glucose)

Hydrolysis      Condensation

Glucose units (monomers)

**Figure 40.2 Hydrolysis of starch.** Starch is a polysaccharide composed of hundreds of glucose subunits. The enzyme amylase hydrolyzes the glycosidic bonds between these subunits. These smaller components diffuse more easily and can be more easily transported into the cell.

short-chain polysaccharides. These smaller carbohydrate molecules can then diffuse more rapidly through the medium and be transported more easily into the cell. These small carbohydrates are excellent sources of carbon and feed directly into the respiratory pathways of bacteria.

Starch agar is made by adding soluble starch to a general growth medium, like nutrient agar. The soluble starch dissolves in the medium and is not visible in the agar. Starch agar plates are inoculated with the organism of interest and then incubated for 24 to 48 hours. Since starch is not visible in the agar, a reagent must be used to identify the presence or absence of starch in the agar. Iodine is the indicator reagent used to identify starch. In the presence of starch, iodine will turn blue to blue-black in color. When iodine is poured onto the starch plate, the agar in the plate will turn blue to blue-black wherever starch is present. If starch has been hydrolyzed, the medium will not change color. A positive test for the production of an extracellular amylase, therefore, is the absence of a blue or blue-black color around the colony after the addition of iodine. If the agar immediately surrounding the colony turns blue or blue-black with the addition of iodine, then starch is still present and the colony is negative for amylase activity.

This medium can be used to differentiate species of *Enterococcus*, *Clostridium*, and *Bacillus*. In the case file for this exercise, while other bacteria such as *Pseudomonas* are more commonly associated with VAP, a wide range of bacteria, including *Bacillus*, have been shown to cause it, and premature infants are at elevated risk. In exercise 44, "Litmus Milk," the case file addresses an outbreak of disease caused by *Bacillus cereus* associated with fried rice. The ability of *Bacillus* to produce an extracellular amylase allows it to grow on rice even when other bacteria cannot.

## ✅ Tips for Success

- Label the starch plates when you pick them up. They look very much like nutrient agar plates, and you don't want to confuse them with other plates at your desk.
- Make short, straight-line streaks. *Bacillus* can quickly overgrow its streak line and may overgrow your other organism and obscure its results.

## Organisms (Broth, Slant, or Plate Cultures)

- *Staphylococcus aureus*
- *Bacillus cereus*

## Materials (per Team of Two or Four)

- 1 starch plate
- Iodine (period 2)
- Loop
- Bunsen burner/microincinerator

## Procedure

### Period 1

1. Use your marker on the bottom of the plate to divide the plate into two sections. Label each section with the appropriate organism name.
2. Use your inoculating loop to aseptically transfer the appropriate organism to each section. Make short streaks.
3. Place the plate in the incubator and incubate for 24 to 48 hours.

### Period 2

1. Remove the plate from the incubator. Record the growth observed on the plate in your lab report.
2. Pour iodine over the plate. Iodine should cover the entire surface of the agar.
3. Record your results. See **figure 40.3** to interpret your results.
4. Discard the plate carefully.

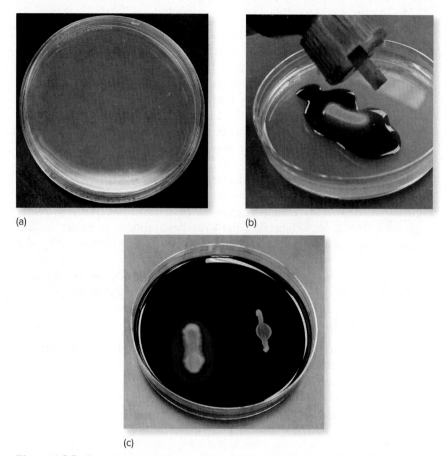

(a)

(b)

(c)

**Figure 40.3 Starch agar plate reactions. (a)** An uninoculated starch plate; **(b)** dropping iodine on the plate. Plate **(c)** has been inoculated, incubated, and treated with iodine. Clearing around the colony in (c) indicates the absence of starch and the presence of extracellular amylase. The blue-black color surrounding the colony on the right indicates starch is still present in the medium. This is a negative amylase test. (a) ©Susan F. Finazzo; (b) ©Steven D. Obenauf; (c) ©Susan F. Finazzo

## Results and Interpretation

Starch agar is a general nutrient agar to which soluble starch has been added. Starch as a polysaccharide composed of many monomers of glucose is potentially a good source of energy for bacteria. Some bacteria produce an exoenzyme, amylase, to break down starch to the level of a monosaccharide or disaccharide. These smaller subunits can easily diffuse through the medium and be transported into the cell. To determine if an organism produces extracellular amylase, the culture is first inoculated onto a starch agar plate. With the addition of iodine, the starch will turn blue-black (figure 40.3). If an organism produces extracellular amylase, then the agar surrounding the colony will be colorless (starch is absent). If the organism does not produce extracellular amylase (amylase-negative), then starch will still be present in the agar surrounding the colony.

# Starch Hydrolysis

## Your Results and Observations

Record your results in the table.

| Organism | Color of agar surrounding the colony after the addition of iodine | Does this organism hydrolyze starch? |
|---|---|---|
| *Staphylococcus aureus* | | |
| *Bacillus cereus* | | |

Draw your results and/or photograph them with your mobile device. Label the diagram appropriately.

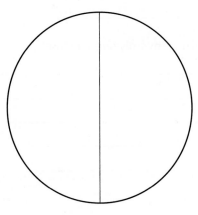

Before the addition of iodine

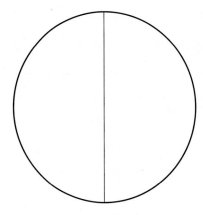

After the addition of iodine

## Interpretation and Questions

1. Why do we add soluble starch to nutrient agar rather than just growing cultures on potatoes?

_____

_____

_____

_____

_____

**2.** Why would the ability to produce extracellular enzymes be a valuable characteristic in the microbial world?

_____

_____

_____

_____

_____

**3.** So far during this course, you have tested for the presence of three extracellular enzymes, DNase (exercise 38), gelatinase (exercise 39), and amylase (exercise 40). If you were a bacterium and had a choice, which enzyme would you produce (select only one)? Why?

_____

_____

_____

_____

_____

**4.** How do you know amylase is an extracellular enzyme?

_____

_____

_____

_____

_____

**5.** Why is amylase less effective at hydrolyzing amylopectin than at hydrolyzing amylose?

_____

_____

_____

_____

_____

# Casein Hydrolysis

The quality control unit at a local milk processing plant had received reports of off-flavor and off-color in the plant's final pasteurized whole milk product. Samples of the suspect product were plated onto various media, including eosin methylene blue (EMB), MacConkey, mannitol salt, nutrient, and casein agar plates. Additionally, swabs were taken at machinery access points throughout the plant. No growth was observed on EMB or MacConkey agar plates. Growth on the other media indicated that the predominant contaminant in the product was a gram-positive, non-mannitol-fermenting, casease-positive coccus. Further testing identified this organism as *Staphylococcus epidermidis*. Samples taken from the plant's machinery indicated that contamination occurred after pasteurization but before the product containers were sealed. The apparatus holding the jets through which pasteurized milk flowed had been serviced earlier in the week. Swabs taken from these jets were also positive for *S. epidermidis*. The apparatus was steam sterilized and the product monitored for further contamination issues.

Source: CDC/Janice Haney Carr

### LEARNING OUTCOMES

At the completion of this exercise, students should be able to

- describe the appearance of a positive and negative casein hydrolysis test result.
- explain the basis for the casein hydrolysis test.
- identify the substrate, enzyme, and enzyme products of the casein hydrolysis test.

## Background

Casein is the large, globular protein that gives milk its white, opaque color. As a protein, casein is a good source of amino acids (carbon and nitrogen). However, casein molecules are too large to be transported across the cell membrane intact. They must be broken down into smaller subunits (amino acids or small peptides) before they can be transported into the cell. Casease is an exoenzyme, secreted outside of the cell, that hydrolyzes casein to amino acids and short peptides. Its action is shown in the following equation:

**Casein + Water → Amino acids**
Casease

Casease production is detected by observing the agar surrounding a colony growing on a casein or milk agar plate. These plates are made by adding skim milk (powder or liquid) to a general growth agar such as trypticase soy agar. The uninoculated plates are opaque and white. If an organism produces and secretes casease, then casein in the agar surrounding the colony will be hydrolyzed and a clear zone will surround the colony. If the organism does not produce casease, then the colony growth will appear on a white background **(figure 41.1)**.

 **Tips for Success**

- Label the casein/milk plate when you pick it up. You don't want to confuse it with other plates at your desk.
- Make short, straight streaks. Some bacteria can quickly overgrow the plate and may obscure the results.

## Organisms (Slant or Broth Cultures)

- *Staphylococcus aureus*
- *Escherichia coli*

## Materials (per Team of Two or Four)

- 1 casein/milk plate
- Loop and Bunsen burner
- Microincinerator

## Procedure

### Period 1

1. Use your marker on the bottom of the plate to divide the plate into two sections. Label each section with the appropriate organism name.
2. Use your inoculating loop to aseptically transfer the appropriate organism to each section. Make short streaks.
3. Place the plate in the incubator and incubate for 24 to 48 hours.

### Period 2

1. Remove the plate from the incubator.
2. Record your results. See figure 41.1 to interpret your results.

## Results and Interpretation

Casein is the protein that gives milk its white color. Casein or milk agar plates are general nutrient agars to which evaporated skim milk is added. Some organisms can produce the enzyme casease, a protease exoenzyme. If the enzyme is produced and secreted into the agar, casein is broken down into smaller peptides and amino acids. Since the protein is no longer structurally intact, the white color disappears. A positive casease reaction on a milk agar plate is a clearing around the colony (figure 41.1*b*). The agar surrounding a colony in which the organism does not produce casease will appear white.

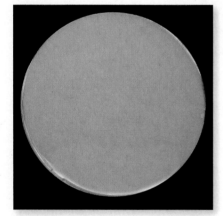
(a)

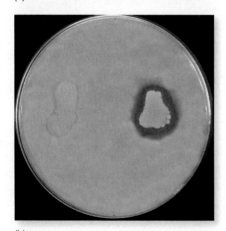
(b)

**Figure 41.1 Hydrolysis of casein.** Plate **(a)** is an uninoculated casein plate. The colony to the right on plate **(b)** is positive for casease (note the zone of clearing around the colony). The colony on the left is negative for casease activity. (a-b) ©Susan F. Finazzo

# Casein Hydrolysis

## Your Results and Observations

Record your results in the table.

| Organism | Changes in the agar surrounding the colony after incubation | Does this organism hydrolyze casein? |
|---|---|---|
| Staphylococcus aureus | | |
| Escherichia coli | | |

Draw your results and/or photograph them with your mobile device. Label the diagram appropriately.

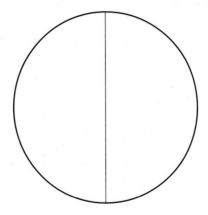

## Interpretation and Questions

**1.** What are the products of casease?

_____

_____

_____

**2.** From the case file and previous exercises, what would growth on EMB and MacConkey agars suggest?

_____

_____

_____

# NOTES

# Phenol Red Broth

## CASE FILE

Over the Memorial Day holiday weekend, 22 patients visited the
emergency room of a regional hospital in Maine, complaining
of flulike symptoms, fever, nausea, muscle aches, weakness,
headache, loss of balance, and altered mental status (four
patients). Eight of the patients, because of preexisting or
chronic conditions, were admitted to the hospital. Two of the
elderly patients succumbed to the infection. A third patient

©Steven D. Obenauf

miscarried in her 26th week of pregnancy. The remaining patients
were given a prescription for a broad-spectrum antibiotic and released.
Blood work on all of the patients indicated moderately elevated WBC
counts. A similar organism was isolated from all of the blood cultures.
Pulsed-field gel electrophoresis patterns for these isolates were indistin-
guishable from each other. The organism was a gram-positive, non-
spore-forming flagellate. The organism was positive for acid production
with no gas with glucose and lactose fermentation, the tests you will
perform in this exercise. In addition, the organism was beta-hemolytic,
catalase-positive, oxidase-negative, methyl red–positive, hydrogen sulfide–
negative, indole-negative, nitrate reductase–negative, and negative for
the hydrolysis of gelatin, starch, and urea. Further carbohydrate
fermentation studies identified the organism as *Listeria monocytogenes*.

Patients were asked to complete a food log for the days preceding
their hospital visit. Epidemiologists from the local health agency
reviewed these reports and found that 20 of the 22 patients had con-
sumed lobster rolls, a ready-to-eat meat product purchased from a local
supermarket and produced at a regional processing plant. Unopened
lobster roll samples from the supermarket and processing plant were
contaminated with *L. monocytogenes*. The U.S. Department of Agriculture
(USDA) has a "zero tolerance" policy for *Listeria* contamination;
therefore, all products originating in the suspect plant were recalled.

## LEARNING OUTCOMES

At the completion of this exercise, students should be able to

- describe the appearance of tubes with the following
  reactions: acid, acid plus gas, and alkaline.
- explain the function of the Durham tube, peptone, pH
  indicator, and carbohydrate source in a medium.
- explain why incubation time is critical in this experiment.

# Background

Carbohydrates are a large, diverse family of organic molecules that function in cells as structural and energy molecules. Simple sugars (monosaccharides) and disaccharides are used by cells to produce adenosine triphosphate (ATP) through the processes of cellular respiration and fermentation. Before bacteria can utilize exogenous carbohydrates, they must first transport them into the cell. Once in the cell, the carbohydrate is directed to the glycolytic pathway. Glycolysis is a catabolic pathway whose end products include two molecules of both pyruvate and ATP. Under oxygen-limiting conditions, pyruvate is further metabolized to a variety of end products, including various alcohols, gases, and acids. These end products not only are economically significant but also are the basis for several tests that are important in the identification of bacteria.

Phenol red broth is a medium that is commonly used to determine the test organism's ability to utilize a specific sugar. Phenol red broth includes the pH indicator phenol red, the sugar of interest, salts, and peptone (soluble hydrolysate of protein). Tubes of the medium also include a small, inverted test tube called a Durham tube. This exercise uses phenol red glucose broth and phenol red lactose broth.

In practice, a single organism is inoculated into a series of broth tubes. Phenol red glucose and phenol red lactose will be used in this exercise. The tubes are then incubated for 24 hours and results observed. If the organism can transport the sugar into the cell, it is further metabolized by intracellular enzymes. By-products of carbohydrate metabolism include various gases (carbon dioxide [$CO_2$], hydrogen gas [$H_2$], and methane [$CH_4$]), acids, and neutral compounds. Gases that are released from the cells accumulate in the Durham tube. As the volume of gas in the tube increases, it displaces an equal volume of media. A bubble will appear in the top of the Durham tube. This is evidence of fermentation. If acids are produced, they, too, are released into the medium. The pH indicator phenol red is red at neutral pH values. As acids are released into the medium and pH decreases, phenol red turns yellow. A yellow tube indicates acid production and indirectly indicates fermentation has occurred.

As just mentioned, if the organism lacks a transmembrane sugar transporter, then the sugar cannot be metabolized. Organisms that lack this transporter may metabolize the peptone in the medium instead. Peptone as a protein derivative is rich in amino acids. Catabolism of amino acids results in the release of ammonia into the medium. Ammonia increases the pH of the medium. Under alkaline conditions, phenol red turns a bright fuchsia red color. A bright fuchsia or magenta red tube is the result of protein, not carbohydrate, metabolism.

Incubation time for this test is **critical.** Inoculated tubes should be interpreted at 24 hours of incubation, no longer. Extended incubation can lead to reversion and inconclusive test results. Bacteria preferentially metabolize carbohydrates. Once carbohydrate supplies have been exhausted, the organism will catabolize any other available nutrient in the environment. In this case, if possible, they catabolize peptone. The breakdown of amino acids in peptone and the release of ammonia cause an increase in the pH of the broth and a change in the phenol red indicator. A tube that is yellow at 24 hours can revert and appear red or fuchsia or yellow with red on the top at 48 hours of incubation. This is called reversion. Unfortunately, most microbiology laboratory courses meet only once or twice a week. To minimize the chance of reversion, your instructor or lab staff may provide specific directions for you to follow with regard to the incubation and evaluation of this medium.

Sugar fermentation studies are useful for differentiating gram-negative enteric bacteria from each other and from other gram-negative organisms.

## ✔ Tips for Success

- All of these tubes look the same! Label the tubes as you pick them up from the supply table.
- As an indicator for gas production, look for the loss of medium from the Durham tube. Some Durham tubes may have a small bubble of gas in them before inoculation. Therefore, only count gas production when at least 10% of the medium in the Durham tube has been displaced by gas production.
- Incubate your tubes for 24 hours only.
- Compare your results to the incubated control tubes. Some color change can occur with incubation alone.

## Organisms (Slant or Broth Cultures)

- *Staphylococcus aureus*
- *Pseudomonas aeruginosa*
- *Escherichia coli*

## Materials (per Team of Two or Four)

- Inoculating loop
- Bunsen burner/incinerator
- 3 phenol red glucose broth tubes
- 3 phenol red lactose broth tubes

## Procedure

### Period 1

1. Label your tubes G or L as you pick them up from the supply table.
2. Label a set of tubes with each organism's name. One set (glucose broth, lactose broth) of uninoculated tubes per class should be used as a control.
3. Use your inoculating loop to aseptically transfer the appropriate organism to each tube.
4. Place tubes in the 37°C incubator and incubate for 24 hours.

### Period 2

1. Remove the tubes from the incubator and record your results. See **figure 42.1** to interpret your results.

## Results and Interpretation

Sugar fermentation capabilities vary widely among bacteria. Carbohydrate fermentation tests are often some of the first tests done in the identification of an unknown organism. Carbohydrate metabolism generates a number of different products. The pH indicator, in this case phenol red, changes color from red to yellow if acidic by-products are present (figure 42.1*b, c,* and *e*). The Durham tube will trap gas released during some forms of fermentation. As gas accumulates in the tube, it displaces the medium. Gas accumulation appears as a bubble in the top of the tube (figure 42.1*c* and *e*). Positive sugar fermentation tubes will always be yellow. Gas production may or may not be present. If tubes are incubated for an extended time, reversion (figure 42.1*e*) can occur. Reversion occurs when the bacteria have exhausted the available carbohydrate and started breaking down the peptone in the medium. The breakdown of protein releases amines into the medium, which increases pH. The phenol red in the medium will revert from yellow to red. Reversion typically begins with a color change reaction at the top surface of the medium.

Some organisms may not ferment the sugar in question. In this case, the bacteria may catabolize the amino acids in peptone. Protein breakdown increases the pH of the medium and causes the phenol red to assume a magenta or bright red color (figure 42.1*d*). This is an alkaline reaction.

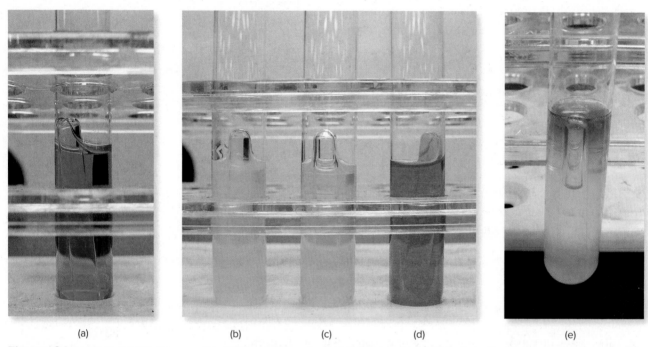

(a)      (b)      (c)      (d)      (e)

**Figure 42.1 Sugar fermentation results. Tube a** is the uninoculated control tube. **Tube b** indicates the production of acid. **Tube c** is acid and gas. **Tube d** is an alkaline reaction. **Tube e** is an example of early reversion. (a-e) ©Steven D. Obenauf

Name _____

Date _____

# Phenol Red Broth

## Your Results and Observations

Record your results after incubation in the table.

| Organism | | Results | | Interpretation |
|---|---|---|---|---|
| | | Color | Gas production | |
| Staphylococcus aureus | Glucose | | | |
| | Lactose | | | |
| Escherichia coli | Glucose | | | |
| | Lactose | | | |
| Pseudomonas aeruginosa | Glucose | | | |
| | Lactose | | | |

## Interpretation and Questions

**1.** Why is this broth made without adding significant amounts of general nutrient extracts?

_____

_____

_____

_____

_____

**2.** One of your fermentation tubes turned a bright red color and plenty of bacterial growth was evident in the tube. What happened? Why is there no bubble in the Durham tube?

_____

_____

_____

_____

_____

**3.** How would you test an organism's ability to ferment another sugar, such as ribulose?

_____

_____

_____

_____

**4.** You are doing sugar fermentation tests. How would you determine that fermentation has occurred?

_____

_____

_____

_____

_____

**5.** Describe the functions of phenol red, the carbohydrate source, Durham tube, and peptone in phenol red fermentation media.

_____

_____

_____

_____

**6.** Why is only a single sugar present in each phenol red tube? Can multiple sugars be tested simultaneously? Why or why not?

_____

_____

_____

_____

# Triple-Sugar Iron Agar

## CASE FILE

A 32-year-old woman visited her dentist for a routine cleaning and exam. X rays revealed an inadequate root filling of the lower right incisor. The old filling was removed and replaced. One week later, the patient complained of pain and tenderness. The gum surrounding the tooth was inflamed and slightly swollen. The root canal was reopened and repacked, and the patient was given a prescription for penicillin. Her symptoms persisted. The root canal was opened again to allow for drainage. The patient was given a prescription for azithromycin. On the fourth day of azithromycin treatment, the patient was pain-free and the root canal was filled permanently. Three days later, the patient reported severe pain, and pus formation was evident. The filling was removed, the wound was irrigated, and samples were sent to a clinical lab for identification using tests that included the ones you will do in this exercise.

©Keith Brofsky/Getty Images RF

Enterobacter cloacae was the only organism isolated from the specimen by the clinical lab. The organism was resistant to penicillin and azithromycin but sensitive to tetracycline and ampicillin. The patient was started on a regimen of ampicillin and was pain free in 3 days. Enterobacter cloacae is a gram-negative facultative enteric bacillus. Although members of this group infrequently cause periodontal infections when they are present, they are associated with persistent apical infections. The presence of enteric bacteria in the oral biota has been attributed to long-term antibiotic therapy, age-related oral problems, contamination of oral solutions, and contamination on the hands of health care professionals.

## LEARNING OUTCOMES

At the completion of this exercise, students should be able to

- describe the appearance of triple-sugar iron agar (TSIA) tubes exhibiting the following reactions: acid, acid plus gas, and alkaline.
- describe the appearance of a TSIA tube in which hydrogen sulfide ($H_2S$) was produced.
- explain the importance of stabbing the butt during inoculation.
- explain why incubation time is critical in this experiment.

# Background

Triple-sugar iron agar (TSIA) is a differential medium used for the differentiation and identification of gram-negative enteric bacteria. The medium tests an organism's ability to ferment selected carbohydrates and to produce hydrogen sulfide. The medium contains three sugars, a pH indicator, two sources of sulfur, an iron salt, and a general nutrient base (peptone, yeast extract). The sugars in TSIA are glucose (0.1%), sucrose (1%), and lactose (1%). Phenol red is the pH indicator. Phenol red is red at neutral pH values, yellow under acidic conditions, and bright red under alkaline conditions. The medium also contains cysteine, a sulfur-containing amino acid, and thiosulfate ($S_2O_3^{2-}$), both of which can serve as sulfur sources and/or electron acceptors. The iron salt is necessary to detect the production of hydrogen sulfide gas ($H_2S$), resulting from the utilization of these sulfur sources.

Fermentation of sugars and the products produced by carbohydrate fermentations were discussed in exercise 40, the phenol red broth carbohydrate exercise. In that exercise, bacteria were inoculated into a broth tube containing only a single sugar. If the organism fermented that sugar, acid or acid and gas were produced, and this was indicated by the change in the color of the medium (red to yellow) and by the displacement of medium from the Durham tube. Unlike the phenol red carbohydrate broth tubes, each tube of TSIA includes three sugars–glucose, lactose, and sucrose. Glucose concentrations are 10 times lower than the concentration of either lactose or sucrose. Organisms growing on the medium typically utilize glucose first. After about 10 hours of incubation, the glucose content of the medium is exhausted. At that point, if possible, the organism switches over to metabolizing sucrose, lactose, or peptone. Color reactions in the medium allow you to differentiate organisms that utilize only glucose, glucose and another sugar, or only peptone. It is possible with this medium to determine if the organism uses sucrose or lactose, but not which of these sugars specifically.

TSIA is dispensed into test tubes to produce a slant with a deep butt (bottom of the tube). The medium is inoculated by stabbing into the butt with an inoculating needle and then streaking down the surface of the slant. This method of inoculation introduces bacteria deep in the butt, below the agar surface under oxygen-limiting conditions that encourage fermentation and then along the slant (aerobic). The slants are then incubated and should be read after 24 hours of incubation. Incubation time for this test, as for other carbohydrate metabolism tests, is **critical.** Inoculated tubes should be interpreted at 24 hours of incubation, no longer. Extended incubation can lead to reversion and inconclusive test results (refer to phenol broth exercise). The lab staff or your instructor may provide you with additional instructions regarding the incubation and evaluation of this medium.

TSIA provides a lot of information about the organism's metabolism. Results include the color reaction of the slant, color reaction of the butt, the production of $H_2S$, and the production of gas **(table 43.1;** also see figure 41.1**).** If an organism can metabolize glucose and another sugar, then the slant and the butt will both give an acid reaction–that is, the slant and butt will be yellow (A/A, or acid slant/acid butt). If this organism also produces gas (G), you may see cracks in the agar or the agar may be displaced or pushed up the tube. If the organism can use only glucose, then the butt will be yellow (acid) and the slant will be red. In this case, the slant initially exhibits an acid reaction (less than 10 hours of incubation), the yellow color then fades, and the slant under the aerobic conditions will turn red. Under aerobic conditions, organisms growing

## Table 43.1 Results and Interpretation of Triple-Sugar Iron Agar

| Results (slant/butt) | Abbreviations* | Interpretation |
|---|---|---|
| No change/no change | NC/NC | No fermentation |
| Red/yellow | K/A | Glucose fermentation plus or minus peptone catabolism |
| Red/yellow plus bubbles | K/A, G | Glucose fermentation; gas produced plus or minus peptone catabolism |
| Yellow/yellow | A/A | Glucose plus lactose or sucrose fermentation |
| Yellow/yellow plus bubbles | A/A, G | Glucose plus lactose or sucrose fermentation; gas produced |
| Yellow/yellow plus black precipitate | A/A, H$_2$S | Glucose plus lactose or sucrose fermentation; hydrogen sulfide produced |
| Red/yellow plus black precipitate | K/A | Glucose fermentation plus or minus peptone catabolism; hydrogen sulfide produced |
| Red/red | K/K | Protein catabolism |

*NC = no change, K = alkaline, A = acid, G = gas, H$_2$S = hydrogen sulfide production.

on the slant may catabolize proteins also, and this produces an alkaline (bright red) reaction. Organisms that do not use any of the sugars in the medium may instead catabolize proteins. Deamination of amino acids releases ammonia into the medium; the pH of the medium increases and the slant and butt will appear bright red. Some organisms utilize sulfur compounds as terminal electron acceptors under oxygen-limiting conditions. Hydrogen sulfide (H$_2$S) is a by-product of these pathways. Hydrogen sulfide is the gas associated with the smell of rotten eggs. This gas is invisible. To detect the production of hydrogen sulfide, an iron salt is included in the medium. This iron salt reacts with hydrogen sulfide to produce iron sulfide, which is visible. This iron sulfide is visible as a black precipitate in the medium. If the black precipitate is present, the organism is classified as H$_2$S-positive.

##  Tips for Success

- Make sure to stab into the butt nearly all the way to the bottom of the tube. Stab through the center of the tube, not along the edge.
- Make sure to streak down the slant.
- Incubate your tubes for 24 hours only.

## Organisms (Slant Cultures)

- *Salmonella typhimurium*
- *Shigella dysenteriae*
- *Alcaligenes faecalis*
- *Escherichia coli*

## Materials (per Team of Two or Four)

- Inoculating loop
- Inoculating needle
- 4 tubes of triple-sugar iron agar (optional–use Kligler Iron Agar)

## Procedure

### Period 1

1. Label your tubes with the appropriate organism's name.
2. Use your needle to aseptically remove a sample of bacteria. Transfer the inoculum to the TSIA tube by stabbing the needle into the butt. Pull the needle back out along that same streak line and then streak down the slant.
3. Place the tubes in the 37°C incubator and incubate for 24 hours.

### Period 2

1. Remove the tubes from the incubator and record your results. See **figure 43.1** to interpret your results.

## Results and Interpretation

Although triple-sugar iron agar can be used to identify $H_2S$-producing bacteria, it is used primarily to study the carbohydrate fermentation capabilities of an organism. The medium contains glucose (0.1%), sucrose (1%), and lactose (1%) as carbohydrate sources. Additionally, the medium contains a pH indicator, two sulfur sources, and an iron salt. Organisms that can utilize only glucose will grow and quickly deplete the glucose in the medium. The acids released from this fermentation will cause the pH indicator in the medium to change from red to yellow. Once fermentation activities slow down, the pH indicator in the slant portion of the tube will revert to a red color within 24 hours. Organisms that utilize only glucose will produce a TSIA tube with an acid butt and alkaline slant **(figure 43.1, tube C)**. Organisms that can use glucose and another sugar (lactose or sucrose) continue to ferment throughout the incubation period and produce a TSIA tube with an acid slant and an acid butt **(figure 43.1, tube B)**. Many organisms using fermentative pathways release gas. If the organism produces gas, the medium may crack or be displaced in the tube (figure 43.1, tubes B and C). Organisms that do not utilize any of the carbohydrates in TSIA can use the peptone in the medium as an energy source. The alkaline by-products of protein catabolism increase the medium's pH and cause the pH indicator to turn magenta or bright red **(figure 43.1, tube A)**.

This medium will also detect hydrogen sulfide production. Hydrogen sulfide is an invisible gas. Iron salts are added to the medium as an indicator for the production of $H_2S$. When $H_2S$ binds to the iron salt in the medium, an insoluble black iron sulfide is formed **(figure 43.1, tube D)**.

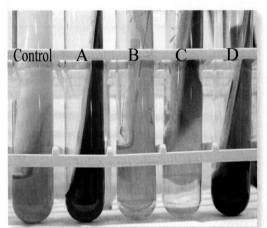

**Figure 43.1 TSIA after 24 hours of incubation.** Notice the orange-red-colored medium of the control tube. **Tube A** is an alkaline reaction, alkaline slant/alkaline butt. **Tube B** is an acid slant/acid butt (A/A) reaction with gas (G) production. **Tube C** has an acid butt and alkaline slant (A/K), indicating only glucose was fermented. This organism also produced gas (G). The reaction in **tube D** is alkaline/acid (K/A) with $H_2S$ production (black precipitate originating in the butt). ©Susan F. Finazzo

Name _____

Date _____

# Triple-Sugar Iron Agar

## Your Results and Observations

Record your results in the table after incubation.

| Organism | Slant color | Butt color | H$_2$S present | Interpretation |
|---|---|---|---|---|
| Salmonella typhimurium | | | | |
| Escherichia coli | | | | |
| Shigella dysenteriae | | | | |
| Alcaligenes faecalis | | | | |

## Interpretation and Questions

1. You are testing an unknown organism from an infected patient. Your TSI tube results show an alkaline slant, an acid butt, and black on the bottom of the tube. Additional testing shows that the organism is indole positive, methyl red positive, and citrate negative. Using the table in exercise 49 (table 49.2), what is the organism?

_____

_____

_____

_____

_____

2. Production of H$_2$S by bacteria growing anaerobically in the intestines is common. Some H$_2$S is absorbed into the circulation and some is expelled as intestinal gas. Less commonly, in certain individuals, H$_2$S-producing bacteria can live in the plaque built up around their teeth. What undesirable effect might that lead to?

_____

_____

_____

_____

**3.** What is the function of iron in this medium?

_____

_____

_____

_____

_____

# Litmus Milk

Fifty-two people sought medical attention at local clinics and emergency rooms in communities near the Raritan River, New Jersey, for gastrointestinal illness. The most commonly seen symptoms described to the nurses taking histories were vomiting, followed several hours later by watery diarrhea and abdominal discomfort. Investigation by the department of health determined that shortly before their symptoms began, all of the patients had eaten at a buffet restaurant and that chicken fried rice was the food most likely associated with illness. Testing of food samples was begun to look for *Bacillus cereus* and other food-borne pathogens.

©Steven D. Obenauf

## LEARNING OUTCOMES

At the completion of this exercise, students should be able to

- describe the two main energy sources that bacteria utilize in milk.
- explain the function of litmus in this medium.
- interpret the appearance of litmus tubes with the following reactions: acid, acid clot, reduction, digestion, and alkaline.

## Background

Litmus milk broth contains the pH indicator litmus in skim milk. Litmus milk is a complex medium that can play a role in identifying a wide range of bacteria. It has two main nutrients that can be metabolized by bacteria: sugar **(lactose)** and protein **(casein).** The indicator in the medium, litmus, can also show a range of reactions from changes in pH to changes in oxidation-reduction status **(figure 44.1).** Below pH 4.5, litmus turns pink. Above pH 8.3, the medium turns blue. Between these values, it is purple.

If the organism ferments lactose, **acid** is produced, turning the litmus pink. Acid production may cause the milk proteins to coagulate or precipitate, forming an acid clot. Tubes demonstrating this reaction will appear pink with white on the bottom or white with a small amount of pink at the top of the tube. Formation of gas during fermentation can produce visible rips or channels in the clot; this is called stormy fermentation. An acid clot is solid. It will not move if the tube is tipped. If the litmus is **reduced** (oxygen is removed), the medium will turn white and will remain fluid.

**Figure 44.1** Examples of the variety of possible reactions in litmus milk.
©Steven D. Obenauf

If the bacterium uses protein as an energy source, different reactions may occur. Partial protein (casein) digestion and the release of ammonia increases pH. This alkaline reaction turns the litmus blue. If the protein is completely broken down and only small peptides and fragments are left, the medium will become less opaque or clear with a brownish tinge. This reaction is called **digestion** or peptonization.

Litmus milk is used for maintaining lactic acid bacteria used in dairy products, for the identification of *Bacillus anthracis*, and for the detection of *Bacillus cereus* in foods. *Bacillus cereus* produces an enterotoxin and can cause food poisoning (intoxication), including one form often associated with fried rice held for hours at warm temperatures.

## ✅ Tips for Success

- Some of the color changes will be quite obvious and others will be more subtle. Using a color control (an uninoculated tube) is **VERY** important in reading this medium.
- Some of the reactions that develop in litmus milk can take up to a week or more to develop. Reincubating and rereading some of your tubes may be needed.
- Handle these tubes carefully after incubation. Shaking or agitating the tubes can make it impossible to interpret the results. Tipping tubes can cause you to spill medium on yourself.

## Organisms (Broth or Slant Cultures)

- *Escherichia coli*
- *Alcaligenes faecalis*
- *Bacillus cereus*

## Materials (per Team of Two or Four)

- 3 tubes of litmus milk
- Inoculating loop
- Bunsen burner/microincinerator

## Procedure

### Period 1

1. Label your tubes with the appropriate organism's name.
2. Inoculate each tube of litmus milk with the appropriate organism.
3. Place your tubes in the incubator.

### Period 2

1. Observe your tubes.
2. Record your results and interpret what they mean.
3. You may need to reincubate your tubes.

## Results and Interpretation

Remember: Some of the color changes will be quite obvious and others will be more subtle. Compare your results to a control tube and the photographs to determine the reaction that has taken place. Compare them to a control tube if needed **(figure 44.2a)**. Be careful not to shake them. If you tip them, do it very carefully. Remember, you are looking at a point in time in a progression of color change, so your tube may not look exactly like any of the photographs. Look for the one that is most similar.

If the bacterium has fermented lactose, **acid** is produced, turning the litmus pink **(figure 44.2b–d)**. Coagulation of milk proteins in an acid environment leads to the production of a clot **(figure 44.2c, d)**. A blue color in the milk **(figure 44.2e)** indicates an **alkaline** reaction, associated with the breakdown of the protein casein. If the milk remains purple and turns white on the bottom, this is indicative of reduction **(figure 44.2f)**. The white will often start on the bottom and eventually almost fill the tube. Finally, if you can see through the milk, this is a sign that the protein is being completely broken down, known as digestion. In some cases it will remain the same color, in some it may turn a brownish, "tealike" color **(figure 44.2g, h)**.

(a)  (b)  (c)  (d)

(e)  (f)  (g)  (h)

**Figure 44.2 Reactions in litmus milk. (a)** Litmus control tube: Compare all others to this tube. **(b)** Acid reaction. **(c)** Acid reaction with an acid clot on the bottom. **(d)** Acid clot with cracks from bubbles ("stormy fermentation"). **(e)** Alkaline reaction. **(f)** Reduction. **(g)** Digestion starting—the milk is beginning to clear. **(h)** Complete digestion—note how you can see through the medium. (a-h) ©Steven D. Obenauf

# NOTES

Name _____

Date _____

# Litmus Milk

## Your Results and Observations

Draw your results or photograph them with your mobile device and describe the result and its interpretation in the table.

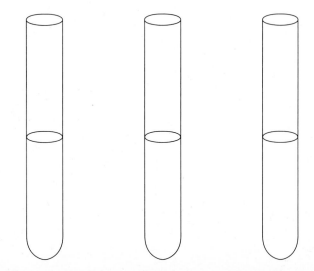

| Organism | Result (appearance/color of the medium) | Interpretation |
|----------|------------------------------------------|----------------|
| *Escherichia coli* | | |
| *Alcaligenes faecalis* | | |
| *Bacillus cereus* | | |
| | | |

## Interpretation and Questions

1. The ability of *Bacillus cereus* to cause disease is associated with the ability it has to produce secreted toxins or exoenzymes. These include emetic toxin (which induces vomiting), hemolysins, enterotoxins, and proteases. Which of these toxins or exoenzymes would be involved in the growth of *B cereus* in litmus milk?

   _____

   _____

   _____

   _____

**2.** As described in the previous question, Bacillus cereus produces emetic toxin, hemolysins, enterotoxins, and proteases. Which of the media used in exercises 14-17 could be used to detect one of these toxins or exoenzymes? Why?

_____

_____

_____

_____

EXERCISE

# 45

# IMViC (Indole, Methyl Red, Voges-Proskauer, Citrate) Reactions

## CASE FILE

Epidemiologists from the regional health care association assessed a cluster of identical *Salmonella enterica* serotype Typhimurium isolates. Isolates were taken from approximately 178 patients residing in five southwestern states. Patient symptoms included diarrhea, abdominal cramping, subjective fever, muscle aches, and bloody stools. In the hospital, patient stool samples were submitted for microbiological testing. A gram-negative enteric organism was suspected. IMViC and differential media (eosin methylene blue [EMB], Endo, and MacConkey agars) were inoculated. *Salmonella enterica* serotype Typhimurium was identified as the causative agent and reported to the Centers for Disease Control and Prevention (CDC).

©Ingram Publishing RF

All of the patients had eaten at local restaurants within the infection period. Three food items—salsa, guacamole, and sliced tomatoes—were common to each case and investigated further. The infection was traced back to tomatoes used to make the salsa. Mixing of produce from different farms at the distributor made it impossible to track the infection to its origination.

Salsa and guacamole food-borne illnesses account for nearly 4% of all food-borne illness cases. These illnesses can lead to hospitalization and, infrequently, death. Salsa and guacamole are made from fresh raw fruits and vegetables. They are not cooked and may spend extended time periods at room temperature before being served. Major sources of contamination include raw vegetables, poor or improper storage, and contamination by food workers.

At the completion of this exercise, students should be able to

- inoculate a sulfur indole motility (SIM) agar tube appropriately.
- describe the appearance of SIM tubes exhibiting the following reactions: indole-positive, indole-negative.
- describe the appearance of a SIM tube in which hydrogen sulfide ($H_2S$) was produced.
- differentiate between motile and nonmotile organisms based on the appearance of the SIM tube.
- describe the appearance of methyl red–positive ($MR^+$) and methyl red–negative ($MR^-$) tubes.
- describe the appearance of Voges-Proskauer-positive ($VP^+$) and Voges-Proskauer-negative ($VP^-$) tubes.
- explain the possible metabolic fates of pyruvate and how the MRVP reactions are used to identify metabolic end products.
- give an example of an $MR^+VP^-$ and $MR^-VP^+$ organism.
- recognize the appearance of a positive citrate reaction.

## Background

The *Enterobacteriaceae* are gram-negative, non-spore-forming, facultative anaerobic, enteric bacilli. The family includes approximately 30 genera and well over 100 species. Although they survive in water and soils, they are most commonly thought of as members of the intestinal flora. The following are some of the genera included in the *Enterobacteriaceae*: *Escherichia, Salmonella, Enterobacter, Shigella, Serratia, Proteus,* and *Yersinia*. Pathogenic members of the enteric group constitute the major causative agents of gastrointestinal infections. They are superb opportunistic pathogens and frequently are associated with nosocomial infections. Clinically, they are very important.

The IMViC tests are actually four separate tests that are used to identify enteric bacteria. IMViC is an acronym that stands for indole, methyl red, Voges-Proskauer, and citrate. The "i" is included for pronunciation purposes. These tests, media, and results are discussed here.

### Indole

SIM medium, along with Kovac's reagent, is used to detect indole production. Indole is produced from the enzymatic hydrolysis of the amino acid tryptophan. Organisms that produce tryptophanase metabolize tryptophan to produce indole, pyruvic acid, and ammonia. SIM is an acryonym for sulfide, indole, and motility. With a single stab inoculation into this semi-solid medium, it is possible to evaluate three different physiological characteristics of the organism.

### SIM and hydrogen sulfide production

The medium contains an iron salt and thiosulfate. When thiosulfate is reduced, the hydrogen sulfide ($H_2S$) gas produced reacts with the iron salt in the medium to form a black iron sulfide precipitate ($H_2S$-positive) **(figure 45.1)**.

## SIM and indole production

SIM contains peptone (semidigested proteins), which provides the organisms with amino acids. Organisms that produce the enzyme tryptophanase hydrolyze tryptophan to produce indole, pyruvic acid, and, through deamination, ammonia. To determine if the organism produces tryptophanase, the medium is first inoculated and incubated. After incubation, Kovac's reagent is added directly to the surface of the agar in the tube. If the reagent turns a cherry-red color (figure 45.1), indole was produced. If the reagent does not change color and remains a pale yellow, the culture is indole-negative **(figure 45.2).**

## SIM and motility determination

This medium is semisolid. The concentration of agar in SIM is about one-fifth the agar concentration of a traditional agar medium. The lower concentration of agar creates a more fluid environment, which allows the bacteria to swim more easily. SIM is inoculated with a needle by stabbing to the bottom of the tube and carefully retracting the needle along the same streak line. **Nonmotile** bacteria will grow in a dense, well-defined line along the stab line (figure 45.2). Motile bacteria will swim away from the stab line. The tube will appear turbid and the stab line will not be visible or distinct.

###  Tips for Success

- Make sure you stab into the tube using the **needle.** You need a steady hand.
- Do not add the Kovac's reagent until after the incubation period.
- The reagent is contained in a glass ampule within the plastic dispenser. Before using the reagent for the first time, the ampule must be broken. To break the ampule, squeeze the plastic dispenser till you hear the glass break. DO NOT remove the ampule from the plastic dispenser.
- The H₂S-positive organisms used in this exercise are also motile. The black precipitate masks the turbidity due to growth of motile organisms.

### Organisms (Slant Cultures)

- *Salmonella typhimurium*
- *Escherichia coli*
- *Klebsiella pneumoniae*

### Materials (per Team of Two)

- Inoculating needle
- 3 tubes of SIM medium

### Procedure

### Period 1

1. Label your tubes with the appropriate organism's name.
2. Use your **needle** to aseptically remove a sample of bacteria. Transfer the inoculum to the SIM tube by stabbing the needle nearly to the bottom of the tube. Pull the needle back out along that same streak line.
3. Place the tubes in the 37°C incubator.

### Period 2

1. Remove the tubes from the incubator and add three drops of the Kovac's reagent. See figures 45.1 and 45.2 to interpret your results.

**Figure 45.1 Hydrogen sulfide production in SIM medium.** The result shown indicates this organism is H₂S-positive and indole-positive (Kovac's reagent is red). ©Steven D. Obenauf

**Figure 45.2 Indole production and motility in SIM medium.** Kovac's reagent has been added to both tubes. The tube on the left is indole-positive (red). The tube on the right is indole-negative. The organism growing in the tube to the right is nonmotile. The organism growing in the tube to the left is motile. ©Steven D. Obenauf

# Results and Interpretation

SIM stands for sulfur indole motility. It is a semisolid medium that is inoculated by stabbing into the agar. SIM medium contains sulfur compounds, which when metabolized release hydrogen sulfide gas. Hydrogen sulfide reacts with iron salts in the medium to produce a black precipitate (figure 45.1). The medium is also used to detect an organism's ability to break down the amino acid tryptophan using the enzyme tryptophanase. The hydrolysis of tryptophan releases indole as a by-product. The presence of indole is detected using Kovac's reagent (figure 45.2). A cherry-red color in the tube on the left indicates the presence of indole (indole-positive) and indirectly indicates the hydrolysis of tryptophan. SIM is also used to assess motility. Motile organisms swim away from the line of inoculation and produce a uniformly turbid tube (figure 45.2, left). Note how the growth of nonmotile organisms is restricted to the line of inoculation (figure 45.2, right).

## Methyl Red and Voges-Proskauer (MRVP) Tests

The methyl red and Voges-Proskauer tests are the M and V tests in the IMViC reactions. They are yet another set of tests that examine carbohydrate metabolism in bacteria. The broth medium contains peptone, glucose, and buffers. All enteric organisms initially metabolize glucose to form pyruvate. Depending on the organism, pyruvate is further metabolized to form either acidic or neutral end products. Some organisms, known as the mixed-acid fermenters, convert pyruvate to a stable acid form, like lactic, acetic, or formic acids **(figure 45.3)**. *Escherichia coli* and *Proteus vulgaris* are both organisms that utilize the mixed-acid pathway. Other organisms use the butylene glycol pathway to convert pyruvate to neutral end products such as acetoin and butanediol, as shown in the following equation:

$$\text{Pyruvic acid} \rightarrow \rightarrow \text{Acetoin (4 C)} + CO_2 + \text{NADH} \rightarrow$$
$$\text{2,3-Butanediol} + NAD^+$$

*Serratia marcescens* and *Enterobacter aerogenes* are examples of organisms that use the butylene glycol pathway. The MRVP medium differentiates between organisms that follow the mixed-acid and butylene glycol pathways.

The MRVP test is done by first inoculating an organism into a single broth tube of MRVP medium. The tube is then incubated. After incubation, half of the medium is dispensed or aliquoted into an empty test tube, so that there are now two tubes of broth. One of these tubes will be used to run the MR test, the other the VP test. You must inoculate and use a single test tube, so that you can compare the MRVP results reliably. The **MR test** is performed by adding several drops of methyl red to one test tube. Methyl red is yellow at pH 6.2 and red at pH 4.4 and below. If the methyl turns red, then the organism has produced a stable acid end product and is considered a mixed-acid fermenter ($MR^+$). If the methyl red remains yellow, then no acid end products were formed ($MR^-$). The second tube is used for the **VP reaction.** VP reagents A and B are added to the tube. If a neutral end product, like acetoin or butanediol, is present, the reagents will turn a red to reddish-brown color ($VP^+$). If no neutral end products are formed, the reagent will remain yellow or yellow-brown in color ($VP^-$). These reactions are typically mutually exclusive, meaning that if an organism is $MR^+$, it is typically $VP^-$, and vice versa.

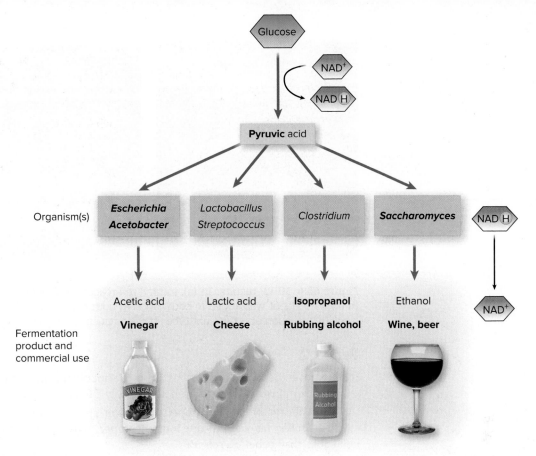

**Figure 45.3 Pathways of pyruvate metabolism.** Pyruvate is an end product of glycolysis and an intermediate product in carbohydrate metabolism. Pyruvate is metabolized by various bacteria to produce either stable acid end products or neutral end products. (a) ©McGraw-Hill Education/Jacques Cornell, photographer; (b) ©Rosemary Calvert/Getty Images RF; (c) ©McGraw-Hill Education/Janette Beckman, photographer; (d) ©Comstock/Getty Images RF

## ✅ Tips for Success

- Label each medium as you pick it up from the supply table. MRVP broth resembles many other broth media.
- Inoculate only a single MRVP tube with each organism. One tube will be used to perform both tests.
- The MR reaction occurs very quickly. The VP reaction may take 30 to 40 minutes to develop. Plan your period 2 activities appropriately.

## Organisms (Slant Cultures)

- *Escherichia coli*
- *Enterobacter aerogenes*

## Materials (per Team of Two or Four)

- Inoculating loop
- 2 tubes of MRVP broth
- Bunsen burner/microincinerator
- (Period 2) Methyl red reagent, VP A, and VP B reagents, 2 empty test tubes

## Procedure

### Period 1

1. Label your tubes with the appropriate organism's name and the medium name. Label the tubes as you pick them up.
2. Use your loop to aseptically transfer the inoculum to the MRVP tube. Repeat with the second organism.
3. Place the tubes in the 37°C incubator.

(a)                                                        (b)

**Figure 45.4 MRVP results.** In **(a)**, methyl red reagent was added to both tubes. In **(b)**, VP reagent was added to both tubes. In both panels, the left tube shows a positive reaction and the right tube shows a negative one. (a-b) ©Steven D. Obenauf

## Period 2

1. Remove the inoculated tubes from the incubator and acquire two empty, nonsterile tubes from the supply bench.

2. Pour/decant or use a dropper to transfer approximately one-half of the *Escherichia coli* culture into an empty, nonsterile culture tube. Label that tube "*E. coli*–VP."

3. Pour/decant or use a dropper to transfer approximately one-half of the *Enterobacter aerogenes* culture into an empty, nonsterile tube. Label that tube "*E. aerogenes*–VP."

4. Add three drops of VP A to each VP tube. Agitate each tube gently. Add six drops of VP B to each VP tube. Agitate gently and allow the tubes to react for 30 minutes. After 30 minutes, compare your results to those shown in **figure 45.4.** The order in which the reagents are added is important. VP A must be added first and VP B second.

5. Add four drops of methyl red to the original MRVP culture tubes of *Enterobacter aerogenes* and *Escherichia coli*. The reaction should be instantaneous. Record your MR results (refer to figure 45.4).

## Results and Interpretation

The methyl red and Voges-Proskauer tests are designed to determine if the end products of carbohydrate metabolism are acidic or neutral. A single tube of MRVP broth is inoculated for each organism to be tested. After incubation, the volume in that tube is split into two separate test tubes. Methyl red is added to one of these test tubes. Methyl red is a pH indicator that will turn red at pH values less than 4.4. If the medium turns red (figure 45.4), the test is MR$^+$, indicating that carbohydrate fermentation produced a stable acid end product. If methyl red did not change colors, then the pH of the medium was 6.2 or higher. A pH close to neutral indicates that no acid end products were produced. The VP reagents (VP A, VP B) are added to the second tube. These reagents react with neutral end products, like acetoin or butanediol, in the medium to produce the brownish-red color. If the tube is VP$^-$, it will remain a yellowish color.

MRVP is a valuable test for distinguishing between members of the *Enterobacteriaceae*. It is particularly useful for identifying *Escherichia coli* and *Enterobacter aerogenes*.

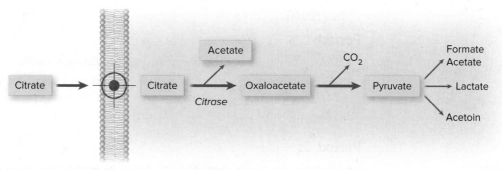

**Figure 45.5 Citrate utilization.** Citrate in the environment must first pass through a transport protein in the cell membrane (bull's-eye). Once in the cytoplasm, citrate is broken down to acetate and oxaloacetate (OAA) by the enzyme citrase. OAA is then converted to pyruvate. The pathway from pyruvate to the final fermentative end product varies with the organism.

## Citrate

The citrate test is the final test in the IMViC series. Simmons citrate is the medium used for this test. The medium contains citrate as the only source of carbon in the medium, ammonium phosphate as the only source of nitrogen, and bromthymol blue as the pH indicator. Citrate is a large, 6-carbon molecule. Citrate is the first intermediate in the Krebs cycle. Some bacteria can also metabolize citrate via an alternate pathway to produce energy **(figure 45.5).**

The ability to use citrate hinges on the organism's ability to transport this large molecule into the cell. Once in the cell, citrase, or citrate lyase, cleaves citrate to form acetate and oxaloacetate. Oxaloacetate, a 4-carbon compound, is then decarboxylated to form pyruvate and carbon dioxide. Pyruvate can then be fermented to produce a variety of acid or neutral end products. Growing organisms also require nitrogen. Bacteria growing on citrate utilize the ammonium phosphate and release ammonia into the medium. Ammonia released from these reactions, along with the carbon dioxide released from the decarboxylation reactions, increases the alkalinity, or pH, of the medium. Bromthymol blue is a pH indicator that is green at neutral pH values and turns bright blue or Prussian blue under alkaline conditions **(figure 45.6).** The pH of uninoculated Simmons citrate agar is 6.9.

A positive citrate test is a Prussian blue slant or visible growth on the slant. Colony growth sometimes has an orange color. This test is useful in the identification of *Enterobacter*, *Klebsiella*, and *Escherichia*. *Enterobacter* and *Klebsiella* are citrate-positive. *E. coli* is citrate-negative.

### ✅ Tips for Success

- Inoculate the slant only. Do not stab the medium. Do not use a heavy inoculum.
- Tubes may need to be inoculated for an entire week.

## Organisms (Slant Cultures)

- *Escherichia coli*
- *Enterobacter aerogenes*

## Materials (per Team of Two or Four)

- Inoculating loop
- 2 slants of Simmons citrate agar
- Bunsen burner/microincinerator

**Figure 45.6 Simmons citrate agar.** The green tube to the left is an uninoculated control slant of Simmons citrate agar. The organism inoculated onto the slant on the right is citrate-positive, as indicated by the Prussian blue color of the medium.
©Steven D. Obenauf

## Procedure

### Period 1

1. Label your tubes with the appropriate organism's name.
2. Use your loop or needle to aseptically remove a small sample of bacteria. Streak the inoculum down the surface of the slant.
3. Place the tubes in the 37°C incubator.

### Period 2

1. Remove the slants from the incubator. A bright blue tube or growth on the slant is a positive reaction for citrate utilization (figure 45.6). If there has been no color reaction or if there is no growth evident on the slant, you may be instructed to reincubate your slants.

## Results and Interpretation

Simmons citrate medium contains citrate as the sole source of combined carbon, ammonium salts as a source of nitrogen, and bromthymol blue as a pH indicator. Only organisms that can import and utilize citrate will grow on this medium. The metabolism of citrate and the ammonium increases the pH of the medium. The bromthymol blue pH indicator changes from green to bright blue under alkaline conditions. Orange colony growth is also a positive citrate reaction.

Name _____

Date _____

# IMViC (Indole, Methyl Red, Voges-Proskauer, and Citrate) Reactions

## Your Results and Observations

Record your results after incubation in the table. You may also wish to use your mobile device to (carefully) photograph your various tubes.

| Organism | SIM | | | MRVP | | Citrate | |
|---|---|---|---|---|---|---|---|
| | Sulfide (+ or −) | Indole (+ or −) | Motility (+ or −) | MR (+ or −) | VP (+ or −) | Slant color | Citrate (+ or −) |
| Salmonella typhimurium | | | | | | | |
| Escherichia coli | | | | | | | |
| Enterobacter aerogenes | | | | | | | |
| Klebsiella pneumoniae | | | | | | | |

## Interpretation and Questions

1. The IMViC tests are based on some enzymatic/physiological process. Name the substrates for the enzymes in the indole, MRVP reactions, and citrate test.

_____

_____

_____

_____

_____

2. Is it possible for an organism to be MR⁺ and VP⁺? Why or why not?

_____

_____

_____

_____

_____

3. Why is it important to inoculate only a single tube per organism when performing the MRVP tests?

_____

_____

_____

_____

4. During lab, your lab partner bumps into you, jostling your arm while you are inoculating your SIM tube. Is it still possible to reliably evaluate the sulfide, indole, and motility results from this tube? Which reactions will still be accurate?

_____

_____

_____

_____

_____

5. The case file describes a cluster of food poisoning incidences. If the causative agent is *Salmonella* spp., then give your prediction of the IMViC test results.

_____

_____

_____

_____

_____

# Biochemical Testing and Diagnosis of Disease

# Urease Test

## CASE FILE

A 53-year-old woman visited her primary care physician complaining of a burning pain in the epigastrium that developed between meals. She also reported episodic occurrences of nausea. The symptoms had exhibited varying severity over the last 6 months. The patient had been self-medicating with over-the-counter drugs, including proton pump inhibitors (lansoprazole [Prevacid], omeprazole [Prilosec]) and antacids (Maalox, Tums), with limited success. The patient's vital signs were normal. She did exhibit some tenderness in the epigastric region. Her doctor's preliminary diagnosis was a gastric ulcer. He referred the patient to the local hospital for a breath test and upper esophagogastroduodenal endoscopy and biopsy.

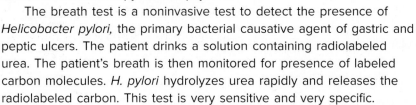

©MedicalRF.com/Getty Images RF

The breath test is a noninvasive test to detect the presence of *Helicobacter pylori,* the primary bacterial causative agent of gastric and peptic ulcers. The patient drinks a solution containing radiolabeled urea. The patient's breath is then monitored for presence of labeled carbon molecules. *H. pylori* hydrolyzes urea rapidly and releases the radiolabeled carbon. This test is very sensitive and very specific.

The patient's breath test was positive for urease activity. The follow-up endoscopic examination identified a single lesion located in the distal portion of the stomach along the lesser curvature. Tissue from the lesion biopsy yielded a positive urease test, and histological examination revealed the presence of a spiral-shaped bacterium, confirming infection with *H. pylori.* The patient was given a 10-day course of amoxicillin and metronidazole for the infection and a proton pump inhibitor to lessen symptoms and promote healing of the gastric mucosa.

## LEARNING OUTCOMES

At the completion of this exercise, students should be able to

- explain the appearance of a positive and negative urease test result.
- explain the basis for the urease test.

# Background

(NH$_2$)$_2$CO (urea) is a nitrogen-containing organic compound. It is a common by-product of amino acid catabolism in animals and constitutes the principal nitrogenous waste product in urine. Urea itself is colorless, odorless, and highly soluble in water.

Urea, with its two amine groups, is a rich source of nitrogen. Some bacteria produce an extracellular urease that hydrolyzes urea to release ammonia and carbon dioxide, as shown in the following equation:

$$(NH_2)_2CO + H_2O \rightarrow CO_2 + 2NH_3$$

Many enteric bacteria can hydrolyze urea but do so slowly. However, some members of the genus *Proteus* are considered rapid urease-positive organisms. The urease test is useful in the identification of *Proteus*, a cause of urinary tract and hospital-acquired infections.

Urea broth is formulated to facilitate the identification of rapid urease-positive organisms. The medium is a highly buffered solution containing urea, a very small amount of yeast extract, and the pH indicator phenol red. Yeast extract provides necessary B vitamins. Phenol red is a pH indicator that is yellow at neutral pH, turns red at alkaline pH values up to pH 8.2, and turns bright pink or fuchsia at pH values greater than 8.2. Urea is the primary source of carbon and nitrogen. The use of urea as the primary carbon source ensures that only urease-positive organisms will grow. The medium is buffered to a neutral pH, and the phenol red will have a yellow-orange color before inoculation. If an organism hydrolyzes urea, the ammonia released from the reaction causes the pH of the medium to increase. The medium will change color from yellow to red to bright pink (fuchsia). Rapid urease-positive organisms metabolize urea quickly and produce ammonia quickly; this overcomes the medium's buffering system, leading to a color change within 24 hours. Organisms that metabolize urea slowly release ammonia more slowly. Slower release and accumulation of ammonia take longer to overcome the buffering system and cause a color change.

## ✔ Tips for Success

- Use a heavy inoculum when preparing your tubes.
- Make sure to compare your tubes to an uninoculated control.

## Organisms (Slant or Broth Cultures)

- *Proteus vulgaris*
- *Escherichia coli*

## Materials (per Team of Two or Four)

- Inoculating loop
- 2 urea broth tubes
- Bunsen burner/microincinerator

## Procedure

### Period 1

1. Label your tubes with the organism's name.
2. Use your inoculating loop to aseptically transfer a heavy inoculum of the appropriate organism to each tube.
3. Place the tubes in the 37°C incubator and incubate for 24 hours.

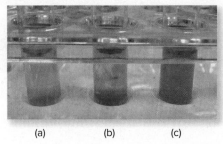

**Figure 46.1 Urease activity. Tube a** is the uninoculated control.
**Tube b** contains a slow urease reaction. **Tube c** is a rapid positive test urease
reaction. (a-c) ©Susan F. Finazzo

## Period 2

1. Remove the tubes from the incubator and record your results. See **figure 46.1** to interpret your results.
2. Some organisms produce a slow positive result. Your instructor may suggest that you reincubate negative test tubes until the next class.

## Results and Interpretation

Urea broth is a highly buffered minimal medium containing urea as the primary carbon source, a small amount of yeast extract, and the pH indicator phenol red. The uninoculated broth (figure 46.1*a*) has a pH of about 6.8 and appears yellow. Organisms that possess the enzyme urease hydrolyze the urea and release ammonia. Some organisms metabolize urea slowly; these produce only a slight color change in the medium (figure 46.1*b*). Rapid urease metabolizers generate more ammonia quickly. The pH increases to 8.2 or higher and the medium turns a bright fuchsia color (figure 46.1*c*).

In addition to urease being part of the diagnosis of *Helicobacter*, as described in the case file, this test is diagnostic for *Proteus* spp.

# NOTES

Name _____

Date _____

# Urease Test

## Your Results and Observations

Record your results after incubation in the table. You may also wish to photograph your tubes.

| Organism | Results (color) | Does this organism produce urease? |
|---|---|---|
| *Proteus vulgaris* | | |
| *Escherichia coli* | | |

## Interpretation and Questions

**1.** Why is it important to limit the available types of nutrients in this broth?

_____

_____

_____

_____

**2.** The case file gives an example of a digestive system infection caused by a urease-positive organism. What other system or systems could commonly be colonized by urease-positive bacteria? Why?

_____

_____

_____

_____

**3.** Urea broth indirectly identifies urease activity by the change in medium pH. How else could urease activity be monitored (consider the case file and the enzyme reaction pathway shown by the equation)?

_____

_____

_____

_____

**4.** You are testing an unknown organism from an infected patient. Your urease tube shows a slow urease reaction. Additional testing shows that the organism is positive for glucose and lactose fermentation, negative for indole, and positive for citrate. Using the table in exercise 49 (49.2), what is the organism?

_____

_____

_____

_____

**5.** Consider the case file. *Helicobacter pylori* metabolizes urea and releases ammonia. How would this affect the microenvironment of the stomach? Is this an important consideration in the treatment of stomach ulcers? Why or why not?

_____

_____

_____

_____

# Nitrate Reduction

## CASE FILE

A 32-year-old female in her 27th week of pregnancy called her gynecologist, complaining of frequent, urgent, and painful urination. When examined at her doctor's office, she was afebrile and her blood pressure, pulse, and respiration were within norms for her condition. She was asked to give a urine sample. The sample was slightly turbid with a visible precipitate. Additionally, the urine tested positive for protein and nitrite, indicating possible bacterial cystitis. This was the third incidence of cystitis during this patient's current gestation. A sample was sent to a clinical lab for bacterial identification and sensitivity testing. The patient was prescribed a 10-day course of nitrofurantoin and asked to return in 2 weeks. Two weeks later, the patient returned. The urinary tract infection had resolved. The doctor prescribed prophylactic low-dosage cephalexin for the remainder of the pregnancy.

©Ingram Publishing RF

The incidence of UTIs in pregnant women approaches 10%. During pregnancy, ureteral dilatation, decreased bladder tone, and increased glycosuria enhance the potential for the development of UTIs. Gram-negative gastrointestinal organisms are the most common causative agents of UTIs; however, group B streptococci can also cause UTIs, which can lead to infections of the vagina. Vaginal group B streptococcal infection can in turn lead to neonatal sepsis, preterm membrane rupture, and preterm delivery. A rare but dangerous possibility is meningitis in the first week of life or after.

## LEARNING OUTCOMES

At the completion of this exercise, students should be able to

- perform a nitrate reduction test.
- explain the function of the reagents used in this exercise.
- identify the enzymes and nitrogen forms in the nitrate reduction pathway.
- explain the significance of denitrification.

# Background

Nitrogen is an essential element required by all living things. Nitrogen is a component of nucleic acids and amino acids and is found in a host of other biological molecules. An organism's survival depends on the organism's ability to find and metabolize suitable forms of nitrogen. The most predominant form of nitrogen on earth is nitrogen gas ($N_2$). Nitrogen gas makes up approximately 80% of the earth's atmosphere. Unfortunately, most organisms, including humans, cannot use this form of nitrogen. We get our nitrogen from the foods that we eat. Nitrogen can also be found in other forms such as ammonium ($NH_4^+$), nitrite ($NO_2^-$), nitrate ($NO_3^-$), and nitrous oxide ($N_2O$). Nitrate is the form preferred by plants.

Bacteria play an essential role in cycling nitrogen through its various forms in the soil, air, and water. Nitrogen-fixing bacteria convert nitrogen gas into a combined form of nitrogen or into an organic form of nitrogen. Members of the genus *Rhizobium* fix nitrogen and form symbiotic relationships with plants. The nitrogenous compounds *Rhizobium* produces help nourish the plant. The plant in turn supplies other nutrients to nourish the bacteria. Nitrogen-fixing bacteria can also be free-living, like members of the genus *Azotobacter*.

Nitrifying bacteria convert ammonium to nitrite and then to nitrate. This is an important process, since plants preferentially use nitrate as a nitrogen source. Unfortunately, nitrate is easily leached into water runoff from agricultural fields. Many fertilizers contain nitrogen in both nitrate and ammonium forms. Nitrate is immediately usable by plants. With time, bacteria in the soil convert ammonium to nitrate.

Denitrifying bacteria reduce nitrate to nitrogen gas or to one of the nitrous oxides ($N_2O$). Agriculturally and economically, this is a very important process. Denitrification decreases the amount of nitrogen available for plants and decreases plant productivity. Since denitrification depletes soil nitrogen content, farmers often have to amend soils with fertilizers. This increases the cost of plant products. Overuse of fertilizers in response to denitrification also has a detrimental effect on the environment. Denitrification occurs under anaerobic conditions. Anaerobic genera such as *Clostridium* perform denitrification as part of their respiratory pathway. When oxygen is limiting, many facultatively anaerobic genera use nitrogenous compounds instead of oxygen as alternate electron acceptors. *Pseudomonas* and many of the gram-negative enteric genera also use nitrogenous compounds in this way. The nitrate test you will perform in this exercise can be part of the process of identifying not only gram-negative bacteria but also some gram-positives such as group B *Streptococcus*.

Some bacteria produce the enzyme nitrate reductase. These organisms use this enzyme to reduce nitrate to nitrite, as shown in the following equation:

$$NO_3 \xrightarrow{\text{Nitrate reductase}} NO_2 \xrightarrow[\text{and other enzymes}]{\text{Nitrite reductase}} N_2O \text{ or } N_2$$

The enzyme nitrate reductase catalyzes the reduction of nitrate ($NO_3$) to nitrite ($NO_2$). Nitrite reductase and several other enzymes catalyze reduction of nitrite ($NO_2$) to nitrogen gas ($N_2$) or nitrous oxides.

Some bacteria not only produce nitrate reductase but also produce nitrite reductase or other nitrite-reducing enzymes. These enzymes then convert nitrite to nitrous oxide or nitrogen gas. In this exercise, you will inoculate nitrate broth with soil and known organisms. Your task will be to determine if nitrate reduction (formation of $NO_2$, $N_2O$, or $N_2$) occurred. After inoculating and incubating your broth tubes, you will add nitrate reagent A (alpha-naphthylamine) and nitrate reagent B (sulfanilic acid)

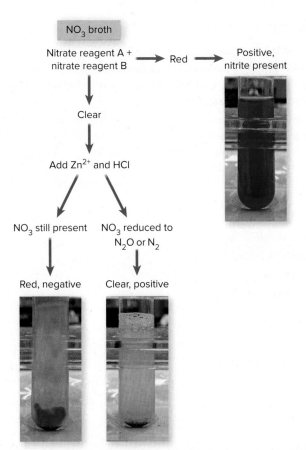

**Figure 47.1 Schematic showing possible outcomes of the nitrate reduction test.** After incubation, reagents nitrate A and B are added to the broth. If the broth immediately turns red, then nitrite is present and the test is positive for nitrate reduction. If the broth remains clear, zinc powder and HCl are added to the broth. These reagents catalyze the reduction of nitrate to nitrite. If nitrate is present in the broth, it will be converted to nitrite. Nitrite will then interact with nitrate reagents A and B and produce a red color. This is a negative test because nitrate was not reduced by the bacteria. If the tube remains clear with the addition of zinc and HCl, then nitrate reduction to a nitrous oxide or nitrogen gas occurred. Since nitrate reduction occurred, this is a positive result. ©Susan F. Finazzo

**(figure 47.1).** These reagents react with **nitrite.** If the broth turns red with the addition of reagents A and B, then nitrite is present and nitrate reduction occurred (positive reaction). But if the broth doesn't react and remains colorless, then the broth either still contains all of the original nitrate or, alternatively, all of the nitrate has been converted to nitrogen gas or nitrous oxide. To distinguish between these two possibilities, you will add a little zinc powder and HCl. Zinc under acidic conditions abiotically catalyzes the conversion of nitrate to nitrite. If nitrate is present, zinc will catalyze the conversion to nitrite. Since nitrate reagent A and nitrate reagent B are already in the tube, the broth will turn red. This is a negative result, since it indicates nitrate was still in the tube. A clear tube after the additions of zinc and HCl indicates that nitrate is not present. Therefore, a clear broth after the addition of zinc and HCl demonstrates that nitrate reduction occurred and the final product was nitrogen gas or a nitrous oxide.

## ✔ Tips for Success

- Do not add the reagents until period 2.
- The color change may not be immediate after adding the reagents A and B or the zinc and HCl. Wait a few minutes before reading and recording the reaction.
- Keep in mind that red color can be both a positive and a negative test result. If red color results from the addition of A and B reagents, then the test is positive for nitrite production (nitrate reduction). If the red color appears in the final step with the addition of zinc and acid, no nitrate was reduced (negative result).

## Organisms

- *Escherichia coli*
- *Pseudomonas aeruginosa*
- *Alcaligenes faecalis*
- *Patient urine sample*

## Materials

- 4 tubes of nitrate broth (with or without Durham tube°)
- Soil
- Loop
- Bunsen burner

- Nitrate reagent A (period 2)
- Nitrate reagent B (period 2)
- HCl (period 2)
- Zinc powder (period 2)

## Procedure

### Period 1

1. Pick up four tubes (or more if you are using both the soil or urine sample) of nitrate broth from the supply table and return to your bench.
2. Use your loop to aseptically inoculate a nitrate broth with *E. coli*. Label the tube with the organism's name and your group's initials.
3. Repeat step 2 with *P. aeruginosa* and *A. faecalis* (or urine sample).
4. Use your loop or a scoopula to add a small amount of soil to a tube.
5. Place the tubes in cans and place the cans in the incubator for 24 to 48 hours.

### Period 2

1. Retrieve your inoculated tubes from the incubator.
2. Add five drops of nitrate reagent A to each tube. Add five drops of nitrate reagent B to each tube. Record your results (red positive). If a tube turned red, you can discard the tube after recording your results. If the tube remained clear, continue on to step 3.
3. Add a small amount of zinc powder and three drops of HCl to the clear tubes. Record your results. If the tube remains clear, it is positive for nitrate reduction. If the tube turns red, it is negative for nitrate reduction (figure 47.1).
4. Record your results and discard all tubes.

°A Durham tube is a small, inverted test tube within the culture tube. It is used to trap gas released during bacterial metabolism.

# Results and Interpretation

Nitrogen is found in various forms, both organic and inorganic. Nitrate is an inorganic form of nitrogen that can be used as a nutrient or as a terminal electron acceptor in the respiratory chain. Nitrate reduction, the first step in denitrification, describes the conversion of nitrate to nitrite. Nitrite can then be further metabolized to various other nitrogenous forms.

Nitrate broth is a minimal medium containing nitrate and a protein digest with no added carbohydrates. After incubation, the nitrate reagents A and B are added to the tube. An immediate color (red) reaction indicates the presence of nitrite (positive nitrate reduction). If the tube does not change color, then there are two possible explanations: Nitrate was not reduced and is still present in the tube, or nitrate was reduced to nitrogen gas or another nitrous oxide not detected by the reagents. The next step in the protocol differentiates between these alternatives. Zinc and HCl are added to the tube. These reagents abiotically catalyze the reduction of nitrate to nitrite. If nitrate was not reduced by the bacteria, these reagents will convert it to nitrite and the tube will turn red. This is a negative test for nitrate reduction. To summarize, there are two possible positive test results for nitrate reduction: a red reaction with the addition of reagents A and B (nitrite produced) and a colorless tube (nitrogen gas or nitrous oxides produced) after the addition of zinc and HCl. A negative test is a red-colored tube with the addition of zinc and HCl.

Nitrate reduction is a characteristic of members of the *Enterobacteriaceae*.

### Table 47.1 Summary of Nitrate Reduction Results

| | Reagents | | | |
|---|---|---|---|---|
| | **A + B** | | **Zinc + HCl** | |
| Color reaction | Red | Colorless | Red | Colorless |
| Nitrogen form in tube | Nitrite ($NO_2$) | Nitrate ($NO_3$) or nitrous oxide | Nitrate ($NO_3$) | Nitrous oxides or $N_2$ |
| Reaction (positive or negative for nitrate reduction) | Positive | | Negative | Positive |

# NOTES

Name _____

Date _____

# Nitrate Reduction

## Your Results and Observations

Record your results in the table. You may also photograph your tubes with your mobile device.

| Organism or sample | Reaction after A + B reagents | Reaction after Zn/HCl* | Interpretation of result |
|---|---|---|---|
|  |  |  |  |
|  |  |  |  |
|  |  |  |  |
|  |  |  |  |

*If needed.

## Interpretation and Questions

**1.** What is the function of the zinc and HCl used in this procedure?

_____

_____

_____

_____

**2.** Which type of respiration uses nitrate as a final acceptor?

_____

_____

_____

_____

**3.** There are two possible ways to observe a positive test for nitrate reduction. What are they?

_____

_____

_____

_____

**4.** Why was the presence of nitrite in the patient's urine (case file) a significant finding?

_____

_____

_____

_____

**5.** Why would you expect denitrification to occur in soil samples?

_____

_____

_____

_____

# "Known" Bacteria for Biochemical Testing

### CASE FILE

You are doing an internship in a diagnostic testing laboratory. The supervisor has assigned you to help with the mainte-nance of their bacterial stock cultures. These cultures are used for quality control testing of culture media and various identification methods. Examination shows that some of the agar that organisms are growing on is drying out, and the cultures may no longer be viable. You have been assigned to make fresh stock cultures of a selection of important bacteria.

©Steven D. Obenauf

### LEARNING OUTCOMES

At the completion of this exercise, students should be able to
- determine if growth on an agar slant is sufficient for use.
- detect contamination in an agar slant culture.
- use cultures repeatedly without introducing contamination.

## Background

The slants you will make today will have two functions. Primarily, they will be the source of bacteria for most or all the biochemical tests that you will be doing for the last part of the semester. These slants may be useful as controls for Gram staining or other aspects of identifying your unknowns, since your unknowns come from the following list of organisms.

### ✓ Tips for Success

- One group in the lab (or the instructor) should keep a set of the 15 broth cultures in the refrigerator or at room temperature in case some of the slants don't grow and need to be reinoculated.
- *Escherichia coli* is used for many of the tests you will be doing. You may need to make a second slant of this organism as a backup for when the first one gets used up.
- Use good aseptic technique. Make sure to flame thoroughly between cultures.

**Figure 48.1** Uninoculated nutrient agar slant. ©Steven D. Obenauf

## Organisms (in Broth Cultures)

- *Escherichia coli* (#1)
- *Enterobacter aerogenes* (#3)
- *Serratia marcescens* (#5)
- *Alcaligenes faecalis* (#7)
- *Enterococcus faecalis* (#12)
- *Proteus vulgaris* (#16)
- *Salmonella typhimurium* (#20)
- *Staphylococcus epidermidis* (#22)

- *Bacillus cereus* (#2)
- *Micrococcus luteus* (#4)
- *Staphylococcus aureus* (#6)
- *Bacillus megaterium* (#8)
- *Klebsiella pneumoniae* (#13)
- *Pseudomonas aeruginosa* (#17)
- *Shigella dysenteriae* (#21)
- *Corynebacterium xerosis* (#11)

## Materials (per Team of Four)

- 16 nutrient agar slants
- Test tube rack
- Bunsen burner/microincinerator
- Inoculating loop

## Procedure

### Period 1

1. Obtain 16 nutrient agar slants (**figure 48.1**), place them in a rack, and label each with the name or number of the bacterium it will contain.
2. Use your inoculating loop to make a slant culture of each of the 16 organisms.
3. Place them in a large container in the incubator.

### Period 2

1. Check all of your slants for growth and possible contamination.
2. Reinoculate any slants (from your original broth culture) that did not grow or were contaminated.
3. Keep your tubes for future use.
4. There is no lab report for this lab.

## Results and Interpretation

Check each slant for sufficient growth. If your loop did not stay in good contact with the surface of the agar, you may have a short line of growth that will not have enough bacteria to use for multiple exercises. Make a new slant. Also check the growth for contamination. All of the growth should have the same color and morphology (see **figure 48.2** for an example). If you see a mixture of colors (e.g., some white, some yellow) or opacity (some clear, some opaque white), make a new slant.

**Figure 48.2** Growth of *Bacillus.* ©Steven D. Obenauf

# Identification of Unknown Bacteria

## CASE FILE

A 5-year-old male is brought to the emergency room with an elevated temperature of 39.1°C (102.4°F), chest pain, and breathing difficulty. Physical examination reveals crackles and diminished lung sounds over the left lung field. His history indicates that he has not received routine childhood vaccinations. Initial laboratory testing shows an elevated white blood cell count with 89% neutrophils and elevated C-reactive protein. Staining of sputum shows both gram-positive and gram-negative bacteria, indicative of a mixed bacterial infection. Additional testing is undertaken to identify each bacterium.

©Steven D. Obenauf

## LEARNING OUTCOMES

At the completion of this exercise, students should be able to

- isolate bacteria from a mixture.
- determine the morphology and Gram reaction of an unknown bacterium.
- determine appropriate additional tests needed after initial staining and fermentation results are obtained.
- interpret the results of multiple biochemical tests and apply them to identification of an unknown bacterium.

## Background

In this series of labs, you will be separating and identifying two bacteria that are mixed together in your unknown tube. Each student will do an unknown. This is an individual process rather than a group one. To identify your bacteria once you have isolated them in pure culture, you will perform a Gram stain and a number of biochemical tests. You will be doing this over a number of lab periods.

## ✓ Tips for Success

- Good aseptic technique is the key to success.
- The first two steps (isolation and Gram staining) take the longest. Don't rush. The biochemical testing usually goes more quickly. A mistake in the beginning will lead you down the wrong path.

## Organisms (in Broth Cultures)

- 1 unknown tube (each has a different number)

## Materials

- 1 nutrient agar plate (period 1)
- 2 nutrient agar slants (possibly additional agar plates) (period 2)
- Additional media that you may use depending on which unknowns you might have (no student will use all of these media): phenol red glucose broth, phenol red lactose broth, sulfur indole motility (SIM) agar, methyl red and Voges-Proskauer (MRVP) broth, citrate slant, triple-sugar iron agar (TSIA) slant, litmus milk, starch plate, casein plate, gelatin agar, nitrate broth or urease broth (period 3 until completion)

## Procedure

### Period 1

1. Pick up one of the tubes of unknowns. **Unless otherwise indicated by your professor, each numbered tube contains two different bacteria.**
2. **Record the tube number** on the lab report page and on your professor's sign-out sheet.
3. Make an **isolation streak plate** (or two if instructed by your professor). Refer to exercise 3 for the isolation streak plate technique.
4. Put your plate in the incubator and your unknown tube in the refrigerator (you will probably need to use it again).
5. Use good aseptic technique, or you may end up with three bacteria instead of two! That would probably be bad. The bacteria in your unknown tube will be two from the following list: *Alcaligenes faecalis, Bacillus cereus, Bacillus megaterium, Corynebacterium xerosis, Enterobacter aerogenes, Enterococcus faecalis, Escherichia coli, Klebsiella pneumoniae, Micrococcus luteus, Proteus vulgaris, Pseudomonas aeruginosa, Serratia marcescens, Salmonella typhimurium, Shigella dysenteriae, Staphylococcus aureus,* and *Staphylococcus epidermidis.*

### Period 2

1. Check your plate for isolated colonies. If you have isolated your unknowns, put each of your two organisms on its own nutrient agar slant. You will use the bacteria on that slant to do your Gram stain (exercise 11) and biochemical testing (exercises 38-47). If you have more than one isolated colony, you can Gram stain the second one.
2. If one or both are not isolated, make a second streak plate.
3. Record the appearance of your organisms on the slant or plate. A few of the possible organisms have a distinct color **(figure 49.1a)**, but most will be white or off-white **(figure 49.1b, c)**. To keep things organized, call your first isolate "A" and your second one "B."

### Period 3

1. If you have one or both of your organisms growing on a slant, Gram stain them. If not, continue isolation. Adding a known gram-positive and known gram-negative to the slide your unknown smear is on (make three smears on the same slide) will help ensure your staining technique is accurate.
2. Record both the shape and Gram reaction of your organisms.

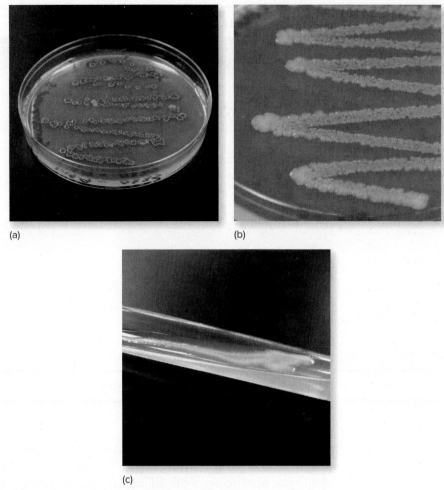

**Figure 49.1** Examples of appearance of colonies. (a-c) ©Steve Obenauf

## Period 4 until completion

1. Based on the Gram reaction of your organisms and their shape, do the biochemical tests needed to identify them. Refer to the information in table 49.2 in the next section to decide which tests you need to do.

2. No one needs to do all the tests. In most cases, a relatively small number of carefully selected tests will be enough to identify your organism. Record your results.

## Results and Interpretation

Your Gram staining and initial results of biochemical testing should be used as the basis for your decision-making process. Once you have narrowed the possible choices to a few organisms, the additional tests you do should be only ones that help you narrow it still further. Tests where all of your possible organisms would give the same result (all positive or all negative) are of no value.

In the example shown in **table 49.1,** the Gram stain showed a gram-negative rod. Fermentation of both glucose and lactose was negative. Only two organisms give those results: *Alcaligenes* and *Pseudomonas*. The tests highlighted in green would be useful to do. The tests that are not highlighted would yield no information.

## Table 49.1 Initial Results and Most Useful Additional Tests to Perform

| Organism | Gram staining | Sugar fermentation | | IMViC tests | | | | |
| | | Lactose | Glucose | Indole | M R | V P | Citrate | H₂S |
|---|---|---|---|---|---|---|---|---|
| *E. coli* | Gram – rod | + (=AG) | + (=AG) | + | + | – | – | – |
| *Enterobacter aerogenes* | Gram – rod | + (=AG) | + (=AG) | – | – | + | + | – |
| *Klebsiella pneumoniae* | Gram – rod | + (=AG) | + (=AG) | – | + or – | + or – | + | – |
| *Proteus vulgaris* | Gram – rod | + or – | + (=AG or A) | + | + | – | – | + |
| *Salmonella typhimurium* | Gram – rod | – or ALK | + (=AG or A) | – | + | – | + or – | + |
| *Shigella dysenteriae* | Gram – rod | – or ALK | + (=AG or A) | + or – | + | – | – | – |
| *Pseudomonas aeruginosa* | Gram – rod | – or ALK | – or ALK | – | – | – | + | – |
| *Alcaligenes faecalis* | Gram – rod | – or ALK | – or ALK | – | – | – | + or – | – |
| *Serratia marcescens* | Gram – rod | – or ALK | + (=AG or A) | – | – | + | + | – |

Abbreviations: A = acid; AG = acid + gas; ALK = alkaline; + or – = this organism could give either result on this test.

Your final decision should be made by comparing all of your results to **table 49.2.** Be aware that some of your test results may be wrong. You may need to do some additional tests to confirm your decision. In the example in **table 49.3,** the Gram stain and sugar fermentations narrowed the probable organisms to two, and the casein hydrolysis made the final determination possible, with the catalase and nitrate tests confirming that determination.

| | | Hydrolysis | | | Oxygen metabolism | | Nitrogen metabolism | | Motility |
|---|---|---|---|---|---|---|---|---|---|
| TSI slant/ butt | Litmus milk | Starch | Casein | Gelatin | Catalase | Oxidase | Nitrate | Urease | (Use SIM or wet mount) |
| ACID/ACID | ACID/CLOT | – | – | – | + | – | + | – | + |
| ACID/ACID | ACID | – | – | – | + | – | + | – | + |
| ACID/ACID | ACID | – | – | – | + | – | + | weak + | – |
| ALK/ACID | ALKALINE | – | + | + | + | – | + | + | + |
| ALK/ACID | ALKALINE | – | – | – | + | – | + | – | + |
| ALK/ACID | REDUCTION | – | – | – | + | – | + | – | – |
| ALK/ALK | DIGESTION | – | + | + | + | + | + | – | + |
| ALK/ALK | ALKALINE | – | – | – | + | + | – | – | + |
| ALK/ACID | REDUCTION | – | + | + | + | – | + | – | + |

## Table 49.2 Organisms and Test Results

| Organism | Gram staining | Sugar fermentation*** | | IMViC tests | | | | |
|---|---|---|---|---|---|---|---|---|
| | | Lactose | Glucose | Indole | M R | V P | Citrate | H₂S |
| E. coli | Gram − rod | + (=AG) | + (=AG) | + | + | − | − | − |
| Enterobacter aerogenes | Gram − rod | + (=AG) | + (=AG) | − | − | + | + | − |
| Klebsiella pneumoniae | Gram − rod | + (=AG) | + (=AG) | − | + or − | + or − | + | − |
| Proteus vulgaris | Gram − rod | + or − | + (=AG or A) | + | + | − | − | + |
| Salmonella typhimurium | Gram − rod | − or ALK | + (=AG or A) | − | + | − | + or − | + |
| Shigella dysenteriae | Gram − rod | − or ALK | + (=AG or A) | + or − | + | − | − | − |
| Pseudomonas aeruginosa | Gram − rod | − or ALK | − or ALK | − | − | − | + | − |
| Alcaligenes faecalis | Gram − rod | − or ALK | − or ALK | − | − | − | + or − | − |
| Serratia marcescens | Gram − rod | − or ALK | + (=AG or A) | − | − | + | + | − |
| Staphylococcus aureus | Gram + cocci | + (=A or AG) | + (=A or AG) | − | + | + or − | − | − |
| Staphylococcus epidermidis | Gram + cocci | + or − | + (=A or AG) | − | − | − | − | − |
| Micrococcus luteus | Gram + cocci | − or ALK | +/− | − | − | − | − | − |
| Enterococcus faecalis | Gram + cocci | + (=A or AG) | + (=A or AG) | − | + | − | − | − |
| Corynebacterium xerosis | Gram + rod | − or ALK | + (=A or AG) | − | − | − | − | − |
| Bacillus cereus° | Gram + rod | − or ALK | + (=A or AG) | − | − | + | + | − |
| Bacillus megaterium | Gram + rod | + (=A or AG) | + (=A or AG) | − | − | − | + | − |

Abbreviations: A = acid; AG = acid + gas; ALK = alkaline; + or − = this organism could give either result on this test.

°Bacillus cereus and Bacillus subtilis cannot be differentiated using biochemical tests. You must look at the morphology in the Gram stain and on your plate or slant and compare it to your "known" cultures.

## Table 49.3 Example of Results Used for Final Identification

| Organism | Gram staining | Sugar fermentation | | IMViC tests | | | | |
|---|---|---|---|---|---|---|---|---|
| | | Lactose | Glucose | Indole | M R | V P | Citrate | H₂S |
| Staphylococcus aureus | Gram + cocci | + (=A or AG) | + (=A or AG) | − | + | + or − | − | − |
| Staphylococcus epidermidis | Gram + cocci | + or − | + (=A or AG) | − | − | − | − | − |
| Micrococcus luteus | Gram + cocci | − or ALK | + /− | − | − | − | − | − |
| Enterococcus faecalis | Gram + cocci | + (=A or AG) | + (=A or AG) | − | + | − | − | − |

Abbreviations: A = acid; AG = acid + gas; ALK = alkaline; + or − = this organism could give either result on this test.

| TSI slant/ butt | Litmus milk** | Hydrolysis | | | Oxygen metabolism | | Nitrogen metabolism | | Motility |
|---|---|---|---|---|---|---|---|---|---|
| | | Starch | Casein | Gelatin | Catalase | Oxidase | Nitrate | Urease | (Use SIM or wet mount) |
| ACID/ACID | ACID + CLOT | − | − | − | + | − | + | − | + |
| ACID/ACID | ACID | − | − | − | + | − | + | − | + |
| ACID/ACID | ACID | − | − | − | + | − | + | weak + | − |
| ALK/ACID | ALKALINE | − | + | + | + | − | + | + | + |
| ALK/ACID | ALKALINE | − | − | − | + | − | + | | + |
| ALK/ACID | REDUCTION | − | − | − | + | − | + | − | − |
| ALK/ALK | DIGESTION | − | + | + | + | + or − | +g | − | + |
| ALK/ALK | ALKALINE | − | − | − | + | + | − | − | + |
| ALK/ACID | REDUCTION | − | + | + | + | − | +g | − | + |
| ACID/ACID | ACID | − | + | + | + | − | + | − | − |
| ALK/ACID | ALKALINE | − | + | − | + | − | + or − | − | − |
| ALK/ALK | ALKALINE | − | + or − | + or − | + | + | + | weak + | − |
| ACID/ACID | ACID | | | | | | | | |
| ALK/ACID | | − | − | − | + | − | + or − | − | − |
| ALK/ACID | ALK or DIGESTION | + | + | + | + | − | + | − | + |
| ACID/ACID | ACID | + | + | + | + | + | + or − | − | + |

**The litmus milk reacts slowly and may change over time. Use it only as a confirmatory test, not as a major part of your decision-making process.

***For sugar fermentation in phenol red broth, a positive reaction is acid production with or without gas; a negative reaction is either no change or alkaline.

g = Nitrate converted to $N_2$ or $N_2O$

| TSI slant/ butt | Litmus milk | Hydrolysis | | | Oxygen metabolism | | Nitrogen metabolism | | Motility |
|---|---|---|---|---|---|---|---|---|---|
| | | Starch | Casein | Gelatin | Catalase | Oxidase | Nitrate | Urease | (Use SIM or wet mount) |
| ACID/ACID | ACID | − | + | + | + | − | + | − | − |
| ALK/ACID | ALKALINE | − | + | − | + | − | − | − | − |
| ALK/ALK | ALKALINE | − | + or − | + or − | + | + | +g | weak + | − |
| ACID/ACID | ACID | − | − | − | − | − | − | − | − |

# NOTES

# Identification of Unknown Bacteria

Organism tube number: _____

## How does it look?

|  | Organism A | Organism B |
|---|---|---|
| Appearance of the colonies |  |  |
| Gram stain: shape |  |  |
| Gram stain: Gram reaction |  |  |

## Initial testing: sugar fermentations

|  | Organism A results | Organism B results |
|---|---|---|
| Phenol red glucose |  |  |
| Phenol red lactose |  |  |

## IMViC tests and others

|  | Organism A results | Organism B results |
|---|---|---|
| Indole |  |  |
| Methyl red |  |  |
| Voges-Proskauer |  |  |
| Citrate |  |  |
| $H_2S$ (on SIM or TSI) |  |  |
| TSI |  |  |
| Casein |  |  |
| Gelatin |  |  |

## Additional tests that are sometimes helpful

| | Organism A results | Organism B results |
|---|---|---|
| Litmus milk | | |
| Starch | | |
| Catalase | | |
| Oxidase | | |
| Urease | | |
| Nitrate | | |
| Motility | | |

Your interpretation:

Organism A is _____

Organism B is _____

Write out your rationale for your choices:

_____

_____

_____

_____

# Index

Note: Page numbers followed by "f" indicate figure, and those followed by "t" indicate table.

Eosin, 67
Eosin methylene blue (EMB) agar, 111–113
Erythrocyte sedimentation rate (ESR), 232
*Escherichia coli*
  biochemical testing and identification, 332,
    334t, 336t
  contamination of water, 111, 112
  genetic engineering, 195
  lactose fermentation, 118
  MRVP tests, 308, 310
  osmotic growth, 128
  *vs. Staphylococcus aureus* (growth comparison),
    128, 130
ESR. *See* Erythrocyte sedimentation rate (ESR)
Ethanol, 78, 309f
*Euglena*, 39, 40, 41f
Eukaryotic microorganisms
  algae and cyanobacteria, 31–35
  fungi, 45–50
  motility, 53–56
  protozoa, 39–42
External ear canal, 61
Eye infections, 77, 202
Eye protection, and safety, 4, 5
Eyewash station, 5

## F

Facultative anaerobes, 134, 135, 136f
Facultative halophile, 128
False motility, 56, 56f
Fastidious organism, 121, 122
FDA, 151
Fecal coliform bacteria, 111–112
Fecal coliforms test, 112
Feet, and safety, 4
Fermentation
  aerotolerance, 134
  identification of unknown bacteria, 336t
  phenol red broth, 288, 289–290
  triple-sugar iron agar, 294, 295t, 296
Fertilizers, 322
Field diaphragm lever, of microscope, 10
Field of view, of microscope, 11
Filamentous colony margin/edge, 22
Filariform larvae, 211
Final dilution, 154, 154f
Fine focus adjustment knob, of
  microscope, 10
Fire blankets, 5, 5f
Fire extinguishers, 5, 5f
First streak area, isolation streak plate, 27
5-Fluorocytosine, 45
Flagella, 33, 34, 34f, 40, 53, 54f
Flagella stain, 59
Flagellated protozoa, 201–202
Flatworms, 210
Flukes, 210
Food(s)
  fungi, 46
  *Listeria* contamination, 287
  nitrogen sources, 322
  tapeworms, 211
  trichinosis, 212
Food-borne infections
  *Bacillus cereus*, 299, 300
  ciguatera toxin, 31
  guacamole, 305
  *Salmonella enterica* serotype
    Typhimurium, 305
  salsa, 305
  sliced tomatoes, 305
  *Staphylococcus aureus*, 25, 105
Food chain, 31, 33
Formic acid, 308
Fossils, of cyanobacteria, 32

Free radical, 161
Frustules, 34
Fungal body, 46. *See also* Hypha
Fungal endocarditis, 45
Fungi, 45–50
  soil, 258–259

## G

*Gambierdiscus toxicus*, 31, 34
Gamma-hemolysis, 99, 100, 101f
Gamma radiation, 161
Gases, as by-products of carbohydrate
  metabolism, 288, 289, 296
Gas gangrene, 134, 134f
Gas jet, safety issues, 3
GasPak jar, 135, 135f
Gastric ulcer, 315
Gastrointestinal illness, 299, 306
Gelatin, 272, 274
Gelatin hydrolysis, 271–274, 335t, 337t
Gel Doc unit, 191
Gel electrophoresis, 287
Gel immunoprecipitation, 225–227
Genetic(s). *See* Applied genetics, and
  transformation
Genetic engineering, 195
Gentamicin, 77, 139, 155, 167,
  169t, 257
Germicidal lamps, 162
*Giardia lamblia*, 204t
Giardiasis, 201
*Giardia* spp., 112, 201, 202f, 203f
Glass, and safety issues, 5
Glomerulonephritis, 100
Gloves, and safety, 4
Glucose
  biochemical testing for identification,
    334t, 336t
  EnteroPluri-Test, 253, 254f
  MRVP tests, 308, 309f
  oxidization, 146
  phenol red, 288
  starch hydrolysis, 277, 278f, 280
  triple-sugar iron agar, 294, 295t, 296
Glucose salts broth (GSB), 123, 124
Glycerol, 259, 262
Glycerol yeast extract agar, 262
Glycocalyx, 33. *See also* Sheath
Glycolysis, 134, 288
Goggles, and ultraviolet light, 162, 162f
Gram, Hans Christian, 77
Gram-negative bacteria
  antibiotic resistance, 172
  cell wall, 78f
  eosin methylene blue agar, 112, 113
  gram staining, 79f, 81, 81f
  MacConkey agar, 118
Gram-positive bacteria
  cells death, 80
  cell wall, 78f, 80
  eosin methylene blue agar, 112,
    113, 113f
  gram staining, 79f, 81, 81f
  MacConkey agar, 117–118
Gram's iodine, 78, 79f
Gram stain, 59, 77–81, 139, 187, 251, 331, 333,
  334t, 336t
Green algae, 33. *See also* Algae
Green fluorescent protein (gfp), 197, 198
Growth, bacterial
  aerotolerance, 133–136
  antimicrobial susceptibility testing,
    167–172
  catalase, 139–141
  measurement of, 122–123

  osmotic pressure and growth, 127–130
  oxidase, 145–147
  plate count, 151–156
Guacamole, 305

## H

Hair, safety, 4
*Halobacterium*, 128
Halophiles, 128
Hand washing, 4, 187
Hanging drop method, for determining
  motility, 53–56
Head trauma, 117
Heart valve replacement, 45
Heat-fixing, of stained slide, 62, 67
*Helicobacter pylori*, 315
Helminths, 210
Hemoglobin, 99
Hemolysins, 99, 100
Hepatitis A virus, 112
Heterocyst, 32–33, 32f
Heterotrophic organisms, 121
Heterotrophs, 47, 121
High concentration of salt, 105–106
High (power) dry objective, 8
Hives, 195
Honey, and botulism, 91
Hookworms, 211–212, 212f
Horizontal gene transfer, 195, 196
Hormogonia, 33
Hospital acquired infections
  DNA fingerprinting, 187, 251
  *Enterobacteriaceae*, 306
  *Enterococcus*, 140
  importance and impact of, 117
  most frequent causes, 53
Hydrochloric acid (HCl), 266, 267, 323, 325,
  325t
Hydrocortisone, 61
Hydrogen peroxide ($H_2O_2$), 139–140
Hydrogen sulfide ($H_2S$)
  EnteroPluri-Test, 254
  IMViC tests, 336t
  SIM medium, 306–307, 307f, 308
  triple-sugar iron agar, 294, 295, 296
Hydrolysis, of casein, 283–284, 284f
Hydrolytic enzymes
  casein hydrolysis, 283–284
  DNase test, 265–267
  gelatin hydrolysis, 271–274
  starch hydrolysis, 277–280
Hyperbaric chamber, 133
Hypha, 46–47, 259
Hypotonic solution, 128, 129f

## I

Identification, of microbes
  DNase test, 265–267
  EnteroPluri-Test, 251–254
  unknown bacteria, 331–337
Immersion objective, 8
Immunoassays, 100
  defined, 237
  rapid, 237–240
Immunodiffusion, 225–227
Immunological memory, 225
Immunoprecipitation testing, 225
Immunosuppression, 271
IMViC reactions, 305–312, 334t, 336t
Incubation time
  phenol red broth, 288
  triple-sugar iron agar, 294
Incubators, 6
Indicators, fecal coliform as, 112